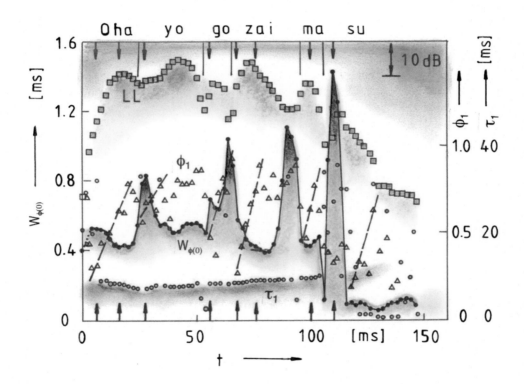

Five temporal factors as a function of time extracted from the running autocorrelation function (ACF) of a continuous speech signal. The times of minimum value, $(\tau_\varepsilon)_{min}$, observed are indicated by arrows. Relative sound pressure level listening level; $\bullet$: $W_{\phi(0)}$; $\circ$: Pitch period τ_1; $\triangle$: Pitch strength ϕ_1, and *dashed line*: development of the pitch strength attaining $\Delta\phi_1/\Delta t$. "Ohayou gozaimasu," single trail.

Signal Processing in Auditory Neuroscience

Signal Processing in Auditory Neuroscience

Temporal and Spatial Features of Sound and Speech

YOICHI ANDO
Professor Emeritus
Kobe University
Kobe, Japan

ELSEVIER

ELSEVIER

3251 Riverport Lane
St. Louis, Missouri 63043

SIGNAL PROCESSING IN AUDITORY NEUROSCIENCE   ISBN: 978-0-128-15938-5

Content Strategist: Melanie Tucker
Content Development Manager: Kathy Padilla
Content Development Specialist: Kristi Anderson
Publishing Services Manager: Deepthi Unni
Project Manager: Nadhiya Sekar
Designer: Gopalakrishnan Venkatraman

Printed in United States of America

Last digit is the print number: 9 8 7 6 5 4 3 2 1

Working together
to grow libraries in
developing countries

www.elsevier.com • www.bookaid.org

Acknowledgments

The author would like to thank: Manfred Schroeder who encouraged him to undertake a series of works since 1970 including concert hall acoustics. M. Schroeder was Director of Drittes Physikalisches Institut, Georg-August-Univ. Goettingen and wrote a recommendation letter to Alexander von Humboldt Foundation, supporting the author to perform the investigations in 1975–77. The author would like to share a memorable episode from a seminar at the institute when after a presentation on subjective preference and the initial time delay gap between the direct sound and the first reflection in relation to the effective duration (τ_e) of ACF, Schroeder pointed out that the value of $1/\tau_1$ could be the pitch (missing fundamental). This moment was the very beginning of investigations into speech by the author. Later, Frau Edith Kuhfuss, Head of Secretaries called the author to her office and explained that the institute may offer a doctorate degree (in physics) if he joined a course for practice. The author's response was that he had just received the degree under Takeshi Itoh, Professor of Waseda University in 1975 and would not have time to perform his own study in parallel.

The author would also like to extend his gratitude toward Hans Werner Strube, who assisted the author with an initial investigation into speech recognition for about 3 weeks in 2011 and gave precise and precious comments. Yoshiharu Soeta, who gave reprints of papers on evoked MEG for each factor extracted from ACF/IACF. Marianne Jõgi, who improved English usage.

Christa Trepte, a long-time friend, who provided room and board in summer of 2011 (Photo P1).

Ando, Y. Architectural Acoustics, Blending Sound Sources, Sound Fields, and Listeners. AIP Press/Springer-Verlag, New York (1998). In part with written permission of the publisher.

Photo P1   Crista Trepte (left) and the author (right) together with Keiko enjoyed tea in the garden on a summer day in 2011, which was arranged by her husband Hainer Trepte during his life. We all became instant friends when we first met at Rohn-sterrassen in 1975. Photographed by Keiko Ando.

Preface

According to neural evidences and cerebral hemisphere specialization related to subjective preference, we have previously proposed a signal-processing model of the auditory brain system. Based on the model we have thoroughly described primary percepts by autocorrelation features as a basis for phonetic and syllabic distinctions. These features have emerged from the theory of subjective preference of the sound field that was originally developed for architectural acoustics. Correlation-based auditory temporal features extracted from monaural autocorrelation are used to predict perceptual attributes such as pitch, timbre, loudness, duration, and coloration of reflected sound. It is worth noticing that, for example, the missing fundamental phenomenon derived from spectral analysis has for a long time remained mysterious, whereas it is described by the factor τ_1 extracted from the running autocorrelation function (ACF) of source signals in very simple terms.

This study investigates the use of features of monaural ACFs for representing phonetic elements (vowels), syllables (/ka/, /sa/, /ta/, /na/, /ha/, /ma/, /ya/, /ra/, /wa/ (CV pairs)), and phrases using a small set of temporal factors extracted from the short-term running ACF. These factors include listening level (loudness), zero-lag ACF peak width $W_{\phi(0)}$ (spectral tilt), τ_1 (voice pitch period), ϕ_1 (voice pitch strength), τ_e (effective duration of the ACF envelope, temporal repetitive continuity/contrast), $(\tau_e)_{min}$ (segment duration), and $\Delta\phi_1/\Delta t$ (the rate of pitch strength change, related to voice pitch attack-decay dynamics). Times at which ACF effective duration τ_e is minimal reflect rapid signal pattern changes that usefully demarcate segmental boundaries. Results suggest that vowels, CV syllables, and phrases can be distinguished on the basis of this ACF-derived feature set, whose neural correlates lie in population-wide distributions of all-order interspike intervals in early auditory stations.

On the other hand, spatial factors extracted from the interaural cross-correlation function (IACF) represent the binaural listening level (LL; mainly binaural loudness), the amplitude of IACF: IACC (subjective diffuseness), binaural time delay τ_{IACC} (localization in the horizontal plane), and the width of IACC: W_{IACC} (apparent source width, ASW). It is obvious that quality of musical sound in a concert hall and speech in a conference room can be comprehensively described by these temporal and spatial factors. Every result obtained from these scientific procedures is always "simple and beautiful" without any complicated mathematical calculation.

Preface

Foreword

It has been more than 30 years since the publication of Yoichi Ando's "Concert Hall Acoustics" (CHA). The current book describes Ando's research since CHA was published, taking a neuroscientific approach to human preferences regarding sound.

The book provides readers with an explanation of the present status of concert hall acoustics, which has been well formulated via both physiological and psychological research. Recent developments in the theoretical understanding of sound preferences can enable new approaches to speech recognition and the design of hearing aids.

As described in Manfred Schroeder's foreword in the original version of CHA, much scientific research in the 1970s and 1980s focused on improving the acoustics of concert halls from the perspective of binaural hearing science. Based on this research, Ando developed the notion of the "four-dimensional world," which describes the most significant factors for concert hall acoustics in terms of preferences for transmitted sound from a source to the listener's ears.

The four dimensions can be summarized as follows:

1. total sound energy,
2. reverberation,
3. delay time of early reflections, and
4. spatial-binaural (interaural) cross-correlation.

These dimensions imply that the qualities of sound fields cannot be represented by a single number. Rather, sound dimensions can be understood from a subjective point of view, such as via listeners' preferences.

The most preferable delay time of early reflections of sound depends on the properties of the sound source, which can be expressed by the autocorrelation functions of a signal.

The autocorrelation function mechanism proposed in the central auditory signal-processing model that provides an estimate of the period of a signal. For example, the fundamental frequency or pitch (the inverse of the estimated period) is thus not necessarily contained in the spectrum of the signal itself. One of the most interesting aspects of Ando's work is the formulation of the relationship between the envelope of the autocorrelation function and the most preferable time delay of an early reflection.

Decaying envelopes are familiar to acousticians in terms of reverberation, which is another dimension in Ando's four-dimensional space.

Responses to a sound source can be formulated by the convolution of the source waveform and the impulse response from the source to the listening position in the sound field.

The impulse response yields the reverberation in the field as an almost exponentially decaying envelope of its waveform.

The decaying property of the reverberation formulates the signal dynamics of traveling sound in the space by convolution, such as the onset and offset portions, and ongoing envelopes.

In contrast, the decaying envelope of the autocorrelation function shows the duration or fluctuation of the periodic properties of sound itself because purely periodic signals are rarely encountered in daily life.

Interestingly, the effective duration, which is when the autocorrelation envelope of a sound decreases to 0.1, referred to as "tau_e" in Ando's terms, provides the most preferable delay time of an early reflection of the sound.

The sound-enhancing interference between direct and reflected sound can be predicted under this criterion, subject to the fusing of direct and reflected sounds into a single sound image. Ando's criterion for the most preferred delay time of an initial reflection, which is formulated by the envelope of the autocorrelation function, makes it possible to design an acoustic space that is adapted to a particular purpose.

Spatial correlations exhibit the periodic characteristics of sound fields, regarding spatial coordinates rather than the time domain. Sinc-functions of spatial coordinates play an important role in spatial cross-correlation functions, as well as those of the time domain for autocorrelation functions. The interaural cross-correlation function of a pair of binaural sounds perceived by a listener, defined in the spatial-time domain, is a new dimension in concert hall acoustics proposed by Ando. Subjective diffuseness is a fundamental notion representing listeners' spatial and temporal impressions in response to sound fields where the spatial correlation, in principle, follows the sinc-function.

Interestingly, dissimilarity of the binaural sound pair is necessary for preferable sound fields. The difference in a pair of binaural signals cannot be perceived in monaural listening, and the dissimilarity is yielded by the time- or phase-difference between the binaural sound pair.

The cross-correlation function is sensitive to such differences, whereas the autocorrelation function is independent of the phase information of signals.

Received sound waves can be represented by the superposition of direct and reverberated sound. The energy ratio of direct sound followed by early reflections and reverberated sound is a conventional parameter representing the sound field from a subjective point of view because the ratio is a function of the distance from the source and its directivity.

In contrast, interaural correlations are determined by the time-structures of binaural sound pairs, which differs from the energy criterion.

Ando developed the theory of acoustic space design, including concert hall design, based on his own neuroscientific research over the past 30 years.

The theoretical and experimental issues are well organized, from introductory matters such as basic terminology and a brief historical review, to explanations of recent research. As such, the book is accessible to a broad audience.

Ando's preference theory focused concert hall acoustics research on the qualities of reverberation and sound fields, such as studies of delay time for early reflections and subjective diffuseness for reverberating sound fields.

Sound is an informative phenomenon and is very much qualitative in nature.

The theoretical understanding of the qualities of sound represents an important subject for sound-based communication, which is an important aspect of daily life cross-culturally.

The research into sound preferences examined in this book involves an exploration of the science of qualities, which is a significant issue for modern science.

Readers will appreciate this well-written book, so that readers find appropriate theme for their future project.

Mikio Tohyama
Fujisawa, Japan

January 2018

Contents

CHAPTER 1

Introduction

When designing enclosures (or any environment) for spoken communication, the acoustic properties of the sound field should be taken into account. This volume describes the human hearing system and the possible auditory mechanisms responsible for the rise of subjective preference. Because subjective preference is a primitive response that steers the judgment and behavior of the organism in the direction of maintaining life, we investigated corresponding cerebral activity in the slow vertex response (SVR) and brain waves in electroencephalogram (EEG) and magnetoencephalography (MEG). The results suggest an auditory signal-processing model that yields primary percepts and a theory of subjective preference for the sound field.[1-4] The temporal and spatial primary percepts may be well described by temporal factors extracted from the running autocorrelation function (ACF) and spatial factors extracted from the interaural cross-correlation function (IACF), respectively. These mechanisms are the bases for automatic speech recognition and the design of hearing aids in the context of this volume.

AUDITORY TEMPORAL AND SPATIAL FACTORS

For the past five decades, we have pursued a theory of architectural acoustics based on acoustics and auditory theory. In the summer of 1971, I visited the III Physics Institute at the University of Goettingen, where Manfred R. Schroeder encouraged me to investigate the aspects of spatial hearing that are most relevant to the design of concert halls. Peter Damaske and I were, at that time, interested in explaining the subjective diffuseness of sound fields using the IACF. The maximum magnitude interaural cross-correlation (IACC) of this function is an indication of the level of subjective diffuseness of a given sound field perceived by an individual due to binaural effects. We reproduced sounds in a room using a multiple-loudspeaker reproduction system and recorded the signals at the two ears of a dummy head.[5] Because the IACC was known to be an important determinant in the horizontal localization of sounds, we also believed it to be significant in subjective

perception of diffuseness. Two years later, in 1974, a comparative study of European concert halls performed by Schroeder, Gottlob, and Siebrasse, showed that the IACC was the most important factor in the incipient subjective preference reactions that established a consensus among individuals.

In early 1975 at Kobe University, we observed a superior sound field with a speech signal that was achieved when adjusting the horizontal direction and the delay time of a single reflection. A loudspeaker in front of a single listener reproduced the direct sound. The angle of the single reflection was about 30 degrees in the horizontal plane measured from the front, the delay time was about 20 ms, and the amplitude was the same as that of the direct sound.[6] These working hypotheses were reconfirmed in the fall of 1975 while the author was an Alexander-von-Humboldt Fellow in Goettingen.[7,8] We were also able to explain the perception of coloration produced by the single reflection in terms of the envelope of the ACF of the source signal.[9]

In 1983, a method of calculating subjective preference at each seat in a concert hall was described by four orthogonal factors of the sound field.[10] Soon after, a concert hall design theory was formulated based on a model of the auditory system. We assumed that some aspects depended on the auditory periphery (the ear), while others depended on processing in the auditory central nervous system.[11] The model takes into account both temporal factors and spatial factors that determine the subjective preference for sound fields.[1] The model consists of a monaural ACF mechanism and an IACF mechanism for binaural processing. These two representations are used to describe monaural temporal and binaural spatial hearing operations that we presume to be taking place at several stations in the auditory pathway, from the auditory brainstem to the hemispheres of the cerebral cortex.

Special attention was given to computing optimal individual preferences by adjusting the weighting coefficients of four orthogonal factors (two temporal factors and two spatial factors), which were used to determine the most preferred seating position for each individual in the room.[2] "Subjective preference" is important to us for philosophical and aesthetic reasons as well as for practical, architectural acoustics reasons. We consider preference as the primitive response of a living creature that directs its judgment and behavior in the pursuit of maintaining life—of body, of mind, and of personality.[12] Thus, neural evidence obtained could be used to identify the auditory system's signal-processing model.

CORRELATION MODEL FOR TEMPORAL AND SPATIAL INFORMATION PROCESSING

To develop a theory of temporal and spatial hearing for room acoustics that is grounded in the human auditory system, we attempted to learn how sounds are represented and processed in the cochlea, the auditory nerve, and in the two cerebral hemispheres. Once effective models of auditory processing are developed, designs for concert halls can proceed in a straightforward fashion, according to guidelines derived from the model.[1] In addition, understanding the basic operations of the auditory system may lead to a new generation of automatic systems for recognizing speech,[4] analyzing music,[2] and identifying environmental noise and its subjective effects.[13] In more general terms, the first book on a brain-grounded theory of temporal and spatial design in architecture and the environment was published in 2016.[14]

It is remarkable that the temporal discharge patterns of neurons at the level of the auditory nerve and brainstem include sufficient information to effectively represent the ACF of an acoustic stimulus. Mechanisms for the neural analysis of interaural time differences through neural temporal cross-correlation operations and for analysis of stimulus periodicities through neural temporal autocorrelations were proposed over half a century ago.[15-17] Since then, many electrophysiological studies based on single neurons and neural populations have more clearly elucidated the neuronal basis for these operations. Binaural cross-correlations are computed by axonal tapped delay transmission lines that feed into neurons in the medial superior nucleus of the auditory brainstem and act as coincidence detectors.[18] If one examines the temporal patterning of discharges in the auditory nerve,[19] one immediately sees the basis for a robust time-domain representation of the acoustic stimulus. Here, the stimulus autocorrelation is represented directly in the interspike interval distribution of the entire population of auditory nerve fibers.[20,21] This autocorrelation-like neural representation subserves the perception of pitch and tonal quality (aspects of timbre based on spectral contour).[22,23]

In our laboratory, we have found neural correlates of spatial hearing in the left and right auditory brainstem responses (ABRs). Here, the maximum neural activity (wave V) corresponds to IACC, that is, the magnitude of the IACF.[24] Also, wave IV for the left and right side brainstem responses ($IV_{l,r}$) nearly corresponds to the sound energies at the right- and left-ear entrances. SVRs are averaged auditory-evoked responses computed from scalp EEG signals. We carried out a series of experiments aimed at developing correlations between brain activity, measurable with the SVR and the EEG, and subjective sound field preference. Subjective sound field preference is well described by four orthogonal acoustic factors, two temporal and two spatial. The two temporal factors are (1) the initial time delay gap between the direct sound and the first reflection (Δt_1), and (2) the subsequent reverberation time (T_{sub}). The two spatial factors are (1) the listening level and (2) the maximum magnitude of the IACF (IACC). The SVR- and EEG-based neural correlates of the two temporal factors are associated with the left hemisphere, whereas the two spatial factors are associated with the right hemisphere.

Higher in the auditory pathway, we reconfirmed by MEG that the left cerebral hemisphere is associated with the delay time of the first reflection Δt_1. We also found that the duration of coherent alpha wave activity (effective duration of the ACF of the MEG response signal) directly corresponds to how much a given stimulus is preferred, that is, the scale value of individual subjective preference.[25,26] The right cerebral hemisphere was activated by the typical spatial factor, that is, the magnitude of IACC.[27] The information corresponding to subjective preference of sound fields was found in the effective duration of the ACF of the alpha waves of both EEG and MEG recordings. The repetitive feature of the alpha wave as measured in its ACF was observed at the preferred condition. This evidence can be pragmatized by applying the basic theory of subjective preference to music and speech signals for each individual's preference.[2]

We also investigated temporal aspects of sensation, such as pitch or the missing fundamental,[28] loudness,[29] timbre,[30] and the duration of sensation.[31] These are well described by the temporal factors extracted from the ACF.[32,33] Remarkably, the temporal factors of sound fields such as Δt_1 and T_{sub} are associated with left hemisphere responses.[24,34-37] Typically, aspects of sound fields involving spatial percepts such as subjective diffuseness[38] and the apparent source width (ASW) have been investigated,[32,39] which are associated with right hemisphere responses.[1,2,26,35,40,41] Tests on dissimilarity judgments were conducted in an existing hall in relation to both temporal and spatial factors.[42]

Features of the IACF correspond to binaural perceptual attributes of binaural listening level, sound direction, ASW, and subjective diffuseness (envelopment).

Neural correlates of ACF-related monaural percepts, which we called "temporal percepts," were observed in electrical and magnetic neural responses over the left cerebral cortical hemisphere, whereas those of binaural IACF-related percepts, which we called "spatial percepts," were observed over the right hemisphere.[3] The correlational features of primary importance to speech recognition lie in the monaural ACF, which could help suppress environmental noise, and in the IACF, which could help suppress unwanted conversations of other people (a spatial attribute that can largely be ignored in telephone communication).

THEORY OF SUBJECTIVE PREFERENCE FOR SOUND FIELDS

Subjective preference of sound fields is deeply related to speech identification and clarity. Preferences present themselves to us as the most primitive responses that steer organisms in the direction of maintaining and propagating life, and are therefore always as relative as they are unique. In humans, preferences are deeply related to the sense of aesthetics. Therefore, making absolute judgments is problematic for reliable results— needless to say, arriving at a result that questions its own objectivity does not constitute reliable information. Instead, preferences should be judged in a relative manner, such as those enabled by the paired comparison test (PCT). It is the simplest and most accurate method, and permits both experienced and inexperienced persons to participate, starting from, for example, 2 years of age when the brain is almost developed. The resulting scale values are appropriate for a wide range of applications. From the results of subjective preference studies with regard to temporal and spatial factors of the sound field, we established the theory of subjective preference.[1,2,10,43]

Over several decades, mainly in the context of architectural acoustics and concert hall design, we have developed a comprehensive theory of auditory signal processing that is based on two internal auditory representations, the monaural ACF and the binaural IACF. The present work extends this general auditory signal processing theory to handle speech distinctions. Features of the two correlation-based representations predict major auditory percepts that are associated with nonspatial temporal sound qualities and spatial attributes, respectively. For example, conference rooms can be designed by applying this preference theory for mainly speech signals to maximize clarity and intelligibility. Facilitating communication of information for the listener is especially important in the presence of other speakers and surrounding noise, both of which are conditions that may separate the perceived signal from the target voice due to the IACF mechanism.

Because individual differences of subjective preference in relation to IACC of spatial factors are very small (nearly everyone has the same basic preferences), we can determine the architectural spatial form of a room by first taking common preferences into account. The temporal factors, which involve successive delays produced by sets of reflective surfaces, are closely related to the dimensions of a specific room/concert hall. These dimensions can be altered to optimize the space due to the minimum effective duration of the running ACF for specific types of sound—music, such as organ music, chamber music, or choral works, or speech.[2]

EXAMPLES OF APPLICATION

The above-mentioned correlation model in the auditory system can be applied to both monaural and binaural speech intelligibility configurations, and to recording identification and simultaneous translation into different languages.

It is often said that children dislike using hearing aids because the models are usually designed by testing only the hearing level. This volume proposes a possible future direction that involves applying the theory of subjective preference to hearing aid design and includes testing and accommodating individual abilities and temporal and spatial percept preferences to increase personal satisfaction.

REFERENCES

1. Ando Y. *Concert Hall Acoustics*. Heidelberg: Springer-Verlag; 1985 [Chapters 1 through 5].
2. Ando Y. *Architectural Acoustics, Blending Sound Sources, Sound Fields, and Listeners*. New York: AIP Press/Springer-Verlag; 1998 [Chapters 1 through 6].
3. Ando Y, ed. *Cariani, Auditory and Visual Sensations*. New York: Springer-Verlag; 2009.
4. Ando Y. Autocorrelation-based features for speech representation. *Acta Acust United Acust*. 2015;101:145−154.
5. Damaske P, Ando Y. Interaural cross-correlation for multichannel loudspeaker reproduction. *Acust*. 1972;27:232−238.
6. Ando Y, Kageyama K. Subjective preference of sound with a single early reflection. *Acustica*. 1977;37:111−117.
7. Ando Y. Subjective preference in relation to objective parameters of music sound fields with a single echo. *J Acoust Soc Am*. 1977;62:1436−1441.
8. Ando Y, Gottlob D. Effects of early multiple reflection on subjective preference judgments on music sound fields. *J Acoust Soc Am*. 1979;65:524−527.

9. Ando Y, Alrutz H. Perception of coloration in sound fields in relation to the autocorrelation function. *J Acoust Soc Am.* 1982;71:616−618.

10. Ando Y. Calculation of subjective preference at each seat in a concert hall. *J Acoust Soc Am.* 1983;74:873−887.

11. Ando Y. Investigations on cerebral hemisphere activities related to subjective preference of the sound field, published for 1983−2003. *J Temporal Des Archit Environ.* 2003;3:2−27.

12. Ando Y. On the temporal design of environments. *J Temporal Des Architect Environ.* 2004;4:2−14. http://www.jtdweb.org/journal/.

13. Soeta Y, Ando Y. *Neurally Based Measurement and Evaluation of Environmental Noise.* Tokyo: Springer; 2015.

14. Ando Y. *Brain-Grounded Theory of Temporal and Spatial Design in Architecture and the Environment.* Tokyo: Springer; 2016.

15. Jeffress LA. A place theory of sound localization. *J Comp Physiol Psychol.* 1948;41:35−39.

16. Licklider JCR. A duplex theory of pitch perception. *Experientia.* 1951;VII:128−134.

17. Cherry EC, Sayers BMA. "Human 'cross-correlator'"—a technique for measuring certain parameters of speech perception. *J Acoust Soc Am.* 1956;28:889−895.

18. Colburn S. Computational models of binaural processing. In: Hawkins H, McMullin T, Popper AN, Fay RR, eds. *Auditory Computation.* New York: Springer-Verlag; 1996.

19. Secker-Walker HE, Searle CL. Time domain analysis of auditory-nerve-fiber firing rates. *J Acoust Soc Am.* 1990;88:1427−1436.

20. Cariani PA, Delgutte B. Neural correlates of the pitch of complex tones. I. Pitch and pitch salience. *J Neurophysiol.* 1996a;76:1698−1716.

21. Cariani PA, Delgutte B. Neural correlates of the pitch of complex tones. II. Pitch shift, pitch ambiguity, phase-invariance, pitch circularity, and the dominance region for pitch. *J Neurophysiol.* 1996b;76:1717−1734.

22. Meddis R, O'Mard L. A unitary model of pitch perception. *J Acoust Soc Am.* 1997;102:1811−1820.

23. Cariani P. Temporal coding of periodicity pitch in the auditory system: an overview. *Neural Plast.* 1999;6:147−172.

24. Ando Y, Yamamoto K, Nagamatsu H, Kang SH. Auditory brainstem response (ABR) in relation to the horizontal angle of sound incidence. *Acoust Lett.* 1991;15:57−64.

25. Soeta Y, Nakagawa S, Tonoike M, Ando Y. Magnetoencephalographic responses corresponding to individual subjective preference of sound fields. *J Sound Vib.* 2002;258:419−428.

26. Soeta Y, Nakagawa S, Tonoike M, Ando Y. Spatial analysis of magnetoencephalographic alpha waves in relation to subjective preference of a sound field. *J Temporal Des Architect Environ.* 2003;3:28−35. http://www.jtdweb.org/Journal/.

27. Sato S, Nishio K, Ando Y. Propagation of alpha waves corresponding to subjective preference from the right hemisphere to the left with change in the IACC of a sound field. *J Temporal Des Architect Environ.* 2003;3:60−69. http://www.jtdweb.org/Journal/.

28. Inoue M, Ando Y, Taguti T. The frequency range applicable to pitch identification based upon the auto-correlation function model. *J Sound Vib.* 2001;241:105−116.

29. Sato S, Kitamura T, Sakai H, Ando Y. The loudness of "complex noise" in relation to the factors extracted from the autocorrelation function. *J Sound Vib.* 2001;241:97−103.

30. Hanada K, Kawai K, Ando Y. *A study of the timbre of an electric guitar sound with distortion. Proceedings of the 3rd International Symposium on Temporal Design.* Guangzhou: J South China University of Technology (Natural Science Edition); 2007:96−99.

31. Saifuddin K, Matsushima T, Ando Y. Duration sensation when listening to pure tone and complex tone. *J Temporal Des Architect Environ.* 2002;2:42−47. http://www.jtdweb.org/Journal/.

32. Ando Y, Sakai H, Sato S. Formulae describing subjective attributes for sound fields based on a model of the auditory-brain system. *J Sound Vib.* 2000;232:101−127.

33. Ando Y. Correlation factors describing primary and spatial sensations of sound fields. *J Sound Vib.* 2002;258:405−417.

34. Ando Y, Kang SH, Morita K. On the relationship between auditory-evoked potential and subjective preference for sound field. *J Acoust Soc Jpn (E).* 1987;8:197−204.

35. Ando Y. Evoked potentials relating to the subjective preference of sound fields. *Acustica.* 1992;76:292−296.

36. Ando Y, Chen C. On the analysis of the autocorrelation function of α-waves on the left and right cerebral hemispheres in relation to the delay time of single sound reflection. *J Architect Plann Environ Eng, AIJ.* 1996;488:67−73.

37. Chen C, Ando Y. On the relationship between the autocorrelation function of the α-waves on the left and right cerebral hemispheres and subjective preference for the reverberation time of music sound field. *J Architect Plann Environ Eng, AIJ.* 1996;489:73−80.

38. Ando Y, Kurihara Y. Nonlinear response in evaluating the subjective diffuseness of sound field. *J Acoust Soc Am.* 1986;80:833−836.

39. Sato S, Ando Y. On the apparent source width (ASW) for bandpass noises related to the IACC and the width of the interaural cross-correlation function (W_{IACC}). *J Acoust Soc Am.* 1999;105:1234.

40. Ando Y, Hosaka I. Hemispheric difference in evoked potentials to spatial sound field stimuli. *J Acoust Soc Am.* 1983;74(S1):S64−S65(A).

41. Ando Y, Kang SH, Nagamatsu H. On the auditory-evoked potentials in relation to the IACC of sound field. *J Acoust Soc Jpn (E).* 1987;8:183−190.

42. Hotehama T, Sato S, Ando Y. Dissimilarity judgments in relation to temporal and spatial factors for the sound fields in an existing hall. *J Sound and Vib.* 2002;258:429−441.

43. Ando Y. Concert hall acoustics based on subjective preference theory. In: Rossing TD, ed. *Springer Handbook of Acoustics.* New York: Springer-Verlag; 2007 [Chapter 10].

CHAPTER 2

Human Hearing System

PHYSICAL SYSTEMS OF THE HUMAN EAR

Head, Pinna, and External Auditory Canal

An acoustic signal is perceived by the ears, in which a sound signal is given by a time sequence. The three-dimensional space is also perceived by the two ears. This is due mainly to the head-related transfer functions $H_{l,r}(r/r_0, \omega)$ between a source point and the two ear entrances, which have directional qualities based on the shape of the head and the pinna system. The directional information is contained in such head-related transfer functions, including the interaural time difference and the interaural amplitude difference.

Fig. 2.1 shows examples of the amplitude of the head-related transfer function $H(\xi,\eta,\omega)$ as parameters of the vertical angle of incidence η ($\xi = 0$), ξ being the horizontal angle. These were measured by the single-pulse method in far-field conditions.[1] The angle $\eta = 0°$, $\xi = 0°$ corresponds to the frontal direction and $\eta = 90°$ to the upper direction. In this situation, the time difference between two ears is $\tau_{IACC} = 0°$, and the only difference is the angular frequency characteristic, ω.

Because the diameter of the external canal is small enough compared to the wavelength below 8 kHz, the transfer function of the entrance canal $E(\omega)$ is independent of the directions in which sound is incident to the human head for the audio frequency range:

$$E(\omega) \sim E_{l,r}(\xi,\eta,\omega) \sim E_{l,r}(\omega) \qquad 2.1$$

Thus, interaction between the sound field in the external canal and that on the outside, including the pinna, is insignificant. The transfer function from the free field to the eardrum can be obtained by multiplying the following two functions.

(1) The sound field from the sound source in the free field to the ear canal entrances, $H_{l,r}(\xi,\eta, \omega)$ and (2) the sound field from the entrance to the eardrum, $E(\omega)$, so

$$H_{l,r}(\xi,\eta, \omega)E(\omega).$$

Measured absolute values of $E(\omega)$ are shown in Fig. 2.2, where variations in the curves obtained by different investigations are caused mainly by the different definitions of the ear-canal entrance point. A typical example

of a transfer function from a sound source in front of the listener to the eardrum is shown in Fig. 2.3. This corresponds to direct sound when the listener is facing a speaker. The transfer functions obtained by three reports[1-3] are not significantly different for frequencies up to 10 kHz.

Eardrum and Bone Chain

Behind the eardrum are the tympanic cavities, which contain the three auditory ossicles: the malleus, incus, and stapes. This area is called the middle ear (Fig. 2.4).

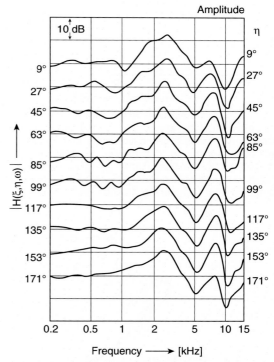

FIG. 2.1 Transfer functions (amplitude) from a free field to the ear-canal entrance as a parameter of the horizontal angle x. (From Mehrgardt S, Mellert V. Transformation characteristics of the external human ear. *J Acoust Soc Am.* 1977;61: 1567–1576.)

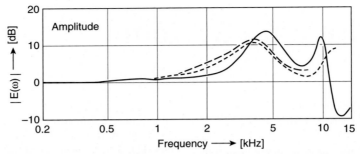

FIG. 2.2 Transfer functions of the ear canal. ([– – – –], From Wiener FM, Ross DA. The pressure distribution in the auditory canal in a progressive sound field. *J Acoust Soc Am*. 1946;18:401–408; [......], From Shaw [1974]; [___], From Mehrgardt S, Mellert V. Transformation characteristics of the external human ear. *J Acoust Soc Am*. 1977;61:1567–1576.)

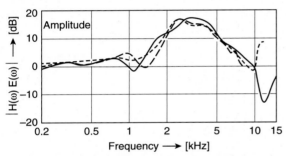

FIG. 2.3 Transfer functions to the eardrum from a sound source in front of the listener. ([– – – –], From Wiener FM, Ross DA. The pressure distribution in the auditory canal in a progressive sound field. *J Acoust Soc Am*. 1946;18: 401–408; [......], From Shaw [1974]; [___], From Mehrgardt S, Mellert V. Transformation characteristics of the external human ear. *J Acoust Soc Am*. 1977;61:1567–1576.)

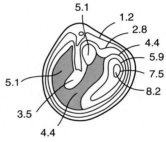

FIG. 2.5 *Contour lines* of equal amplitude of human eardrum vibration at 525 Hz (121 dB SPL). Each value should be multiplied by 10^{-5} cm. (From Tonndorf J, Khanna SM. Tympanic-membrane vibrations in human cadaver ears studied by time-averaged holography. *J Acoust Soc Am*. 1972;52:1221–1233.)

The sound pressure striking the eardrum is transduced into vibration. The middle-ear ossicles transmit the vibration to the cochlea. The vibration pattern of the human eardrum was first measured by Bekesy,[4] who performed a point-by-point examination with an electric capacitive probe. Later, Tonndorf and Khanna[5] measured the vibration pattern by time-averaged holography, which allows perception of finer vibration patterns on the eardrum, as shown in Fig. 2.5. Note that the outline of the malleus is visible in the pattern at the amplitude value of 3.5. The vibration on the malleus is transmitted to the incus and the stapes. The transfer function $C(\omega)$ of the human middle ear (three auditory ossicles) between the sound pressure at the eardrum and the apparent sound pressure on the cochlea is plotted in Fig. 2.6. Data were obtained by Onchi[6] and Rubinstein[7] from cadavers. The values have been rearranged by the Ando.[8]

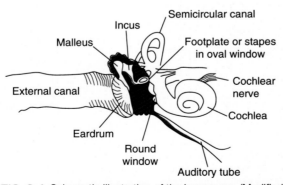

FIG. 2.4 Schematic illustration of the human ear. (Modified from Dorland, W.A.N. (1947). The American Illustrated Medical Dictionary. 24th ed. Saunders, Philadelphia.)

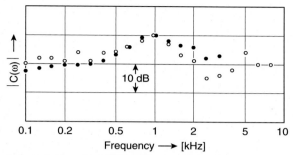

FIG. 2.6 Transfer function (relative amplitude) of the human middle ear between the sound pressure at the eardrum and the apparent pressure on the cochlea. (●), Average value measured (modified from Onchi, 1961); (○), measured value (Rubinstein, 1966.)

The maxima at 1 kHz are adjusted to the same value. Later, Puria, Rosowski, and Peake[9] made measurements by a system that included a hydropressure transducer used in the vestibule as shown in Fig. 2.7. The hydropressure transducer and the microphone with identical sound pressure stimuli in air produced estimates of pressure within 0.5 dB for the range of 50 Hz to 11 kHz. The results at the sound pressure levels of 106, 112, and 118 dB indicating similar values are shown in Fig. 2.8. These results agree well with the data shown in Fig. 2.6, as far as relative behavior is concerned. The transfer function measured at 124 dB showed some signs of nonlinearity, but was consistent with a linear system below approximately 118 dB. The magnitude of the middle-ear pressure gain is about 20 dB in the frequency range from 500 Hz to 2 kHz.

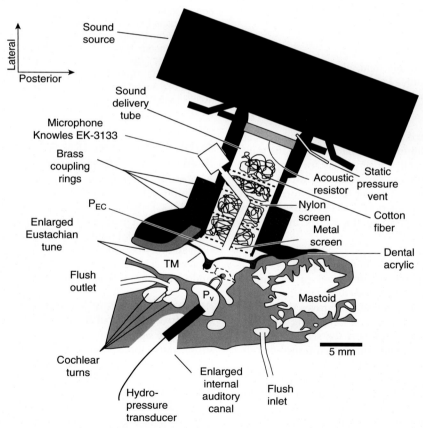

FIG. 2.7 Measurement system of middle ear transfer function. To measure the inner-ear pressure, a hydropressure transducer was placed in the vestibule facing the stapes. In order to ensure that the cochlea remains fluid filled during the measurement, an inlet flush tube was cemented into the superior semicircular canal and an outlet flush tube was cemented into the apical turn of the cochlea. (Puria S, Rosowski JJ, Peake WT. Middle-ear pressure gain in humans: preliminary results. In: Duifhuis H, Horst JW, Dijk P, Netten SM, eds. *Proceedings of the International Symposium on Biophysics of Hair Cell Sensory Systems*. Singapore: World Scientific; 1993:345–351.)

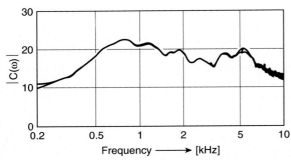

FIG. 2.8 Transfer function of the human middle ear between the sound pressure at the eardrum and the inner-ear pressure. The global behavior is surprisingly similar to that in Fig. 2.6. (Puria S, Rosowski JJ, Peake WT. Middle-ear pressure gain in humans: preliminary results. In: Duifhuis H, Horst JW, Dijk P, Netten SM, eds. *Proceedings of the International Symposium on Biophysics of Hair Cell Sensory Systems*. Singapore: World Scientific; 1993:345–351.)

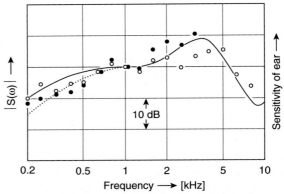

FIG. 2.9 Sensitivity of the human ear to a sound source in front of listeners. (____), Normal hearing threshold (ISO recommendation); (......), reexamined in the low-frequency range. (●, ○), Transformation characteristics between the sound source and the cochlea, S(w) = H(w)E(w)C(w); (●), data obtained from measured values C(w) by Onchi (1961), and (○), from Rubinstein (1966), which are combined with the transfer function H(w)E(w) measured by Mehrgardt and Mellert (1977). (Berger EH. Re-examination of the low-frequency (50–1000 Hz) normal threshold of hearing in free and diffuse sound fields. *J Acoust Soc Am*. 1981;70: 1635–1645.)

For the usual sound field, the transfer function between a sound source located in front of the listener and the cochlea can be represented by

$$S(\omega) = H(0,0,\omega)E(\omega)C(\omega). \qquad (2.1)$$

The values are plotted in Fig. 2.9 based on cadaver data from Onchi[6] and Rubinstein[7]. The pattern of the transfer function agrees with ear sensitivity for people with normal hearing ability, so ear sensitivity can be characterized primarily by the transfer function from the free field to the cochlea Zwislocki[10]. Better agreement can be obtained by reexamining the values in the low-frequency range Berger[11].

The Cochlea

The stapes, which is the smallest bone in the human body, is also the last of the three auditory ossicles. It is connected to the oval window, and drives the fluid in the cochlea, producing a traveling wave along the basilar membrane. The cochlea contains the sensory receptor organ on the basilar membrane, which transforms the fluid vibration into a neural code, as shown in Fig. 2.10. The basilar membrane is so flexible that each section can move independently of the neighboring section. The traveling waves on the basilar membrane observed by Bekesy[4] and shown in Fig. 2.11A and B, are consistent with this representation.

AUTITORY BRAINSTEM RESPONSES IN AUDITORY PATHWAYS

A possible mechanism for a spatial factor is the maximum value of the interaural cross-correlation function (IACC), between sound signals arriving at the two ear entrances for percepts of localization, apparent source width (ASW), and subjective diffuseness. The left and right auditory-brainstem responses (ABRs) were recorded to justify such a mechanism for spatial information Ando, Yamamoto, Nagamatsu and Kang[12].

Auditory-Brainstem Response Recording and Flow of Neural Signals

As a source signal p(t), a short-pulse signal (50 μs) was supplied to a loudspeaker with frequency characteristics for 100 Hz to 10 kHz, ±3 dB. This signal was repeated every 100 ms for 200 seconds (2000 times) to be integrated simultaneously, and left and right ABRs were recorded through electrodes placed on the vertex and left and right mastoids. The distance between loudspeakers and the center of the head was kept at

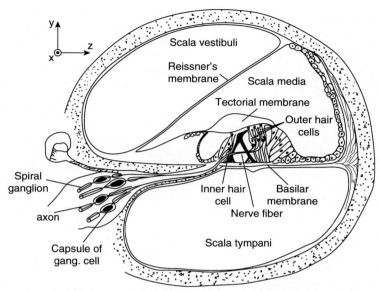

FIG. 2.10 Cross section through the cochlea showing the fluid-filled canals and the basilar membrane supporting hair cells. (Modified from Rasmussen AT. *Outline of Neuro-Anmy.* 3rd ed. Dubuque, IA: William C. Brown; 1943.)

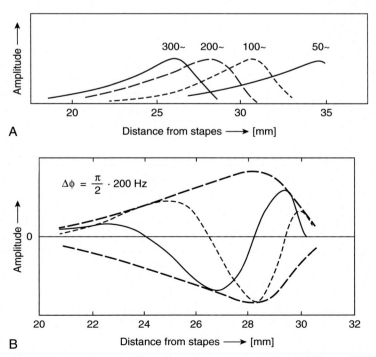

FIG. 2.11 Envelope of the traveling wave (A), and traveling waves on basilar membrane at 200 Hz (B). (Bekesy G. *Experiments in Hearing* [Wever E, transl. and ed.]. New York: McGraw-Hill; 1960.)

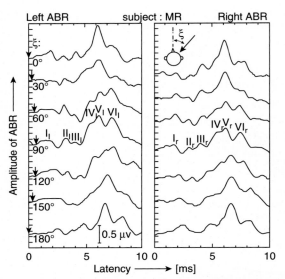

FIG. 2.12 Examples of auditory brainstem response *(ABR)* obtained between the vertex and left and right mastoids, as a function of the latency less than 10 ms and a parameter of the horizontal angle of sound incidence. The abscissa is the latency in the auditory pathways relative to the time when the single pulse arrives at the right ear entrance. *Arrows* indicate the time delay, depending upon the sound source location of the right-hand side of the subject, and the null amplitude of ABR. The wave number is defined by the symbol from $I_{l,r}$ to $VI_{l,r}$, which is reflected by the activity at each nucleus (relay station) in the auditory systems. The suffix signifies the left and right auditory pathway. (Ando, Yamamoto, Nagamatsu and Kang, 1991.)

68 ± 1 cm. The loudspeakers were located on the right-hand side of the subject.

Examples of recorded ABR as a parameter of the horizontal angle of sound incidence of one of four subjects are shown in Fig. 2.12. It is evident that waves I to VI from the vertex and the right mastoid differ in amplitude as indicated by each curve. The ABR data were quite similar among the four subjects who participated, and thus data for the four subjects (23 ± 2 years of age, male) were averaged. As shown in Fig. 2.13A (wave I), of particular interest is the fact that amplitudes from the right, which may correspond to the sound pressure from the source located on the right-hand side, are greater than those from the left; r > l for $\xi = 30° - 130°$ ($P < .01$). This tendency is reversed in wave II as shown in Fig. 2.13B; l > r for $\xi = 60°$ and $90°$ ($P < .05$). The behavior of wave III, shown in Fig. 2.13C, is similar to that in wave I; r > l for $\xi = 30° - 120°$ ($P < .01$). This tendency is again reversed in wave IV as shown in Fig. 2.13D; l > r for $\xi = 60° - 150°$ ($P < .05$), and this is maintained further in wave VI as shown in Fig. 2.13F, even though absolute values are amplified; l > r for $\xi = 60° - 150°$ ($P < .05$). From this evidence, it is likely that the flow of neural signals is interchanged three times between the cochlear nucleus, the superior olivary complex, and the lateral lemniscus as shown in Fig. 2.14 for this spatial information process. The interchanges at the inferior colliculus may be operative for interaural signal processing as discussed later.

In wave V, shown in Fig. 2.13E, such a reversal cannot be seen, and the relative behavior of amplitudes of the left and right are parallel and similar. Thus, these two amplitudes were averaged and plotted in Fig. 2.17 (symbols V). In this figure, the amplitudes of wave IV (left and right, symbols l and r) are also plotted in reference to the ABR amplitudes at frontal sound incidence.

Concerning the latencies of waves I through VI relative to the time when the short pulse was supplied to the loudspeaker, the behaviors indicating relatively short latencies in the range around $\xi = 90°$ were similar (Fig. 2.16). It is remarkable that a significant difference is achieved ($P < .01$) between averaged latencies at $\xi = 90°$, and those at $\xi = 0°$ (or $180°$, i.e., a difference of about 640 µs on average, which corresponds to the interaural time difference of sound incident at $\xi = 90°$). It is most likely that the relative latency at wave III may be reflected in the interaural time difference. No significant differences could be seen between the latencies of the left and right of waves I to IV, as indicated in Fig. 2.16.

Auditory-Brainstem Response Amplitudes in Relation to Interaural Cross Correlation

Fig. 2.17 shows values of the magnitude of interaural cross correlation and the autocorrelation functions (ACFs) at the time origin. These were measured at the two ear entrances of a dummy head as a function of the horizontal angle after passing through the A-weighting networks (nearly characteristic of ear sensitivities). The averaged amplitudes of waves IV (left and right) and averaged amplitudes of wave V, which were both normalized to the amplitudes at the frontal incidence, are shown in Fig. 2.15.

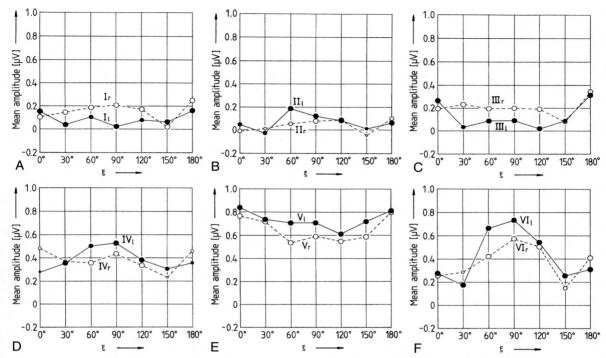

FIG. 2.13 Averaged amplitudes of auditory brainstem responses (ABRs) for four subjects, waves I–VI. Sizes of the *circles* indicate the number of available data from four subjects. *Filled circles*: Left ABRs; *Empty circles*: Right ABRs. (A) Wave I, (B) Wave II, (C) Wave III, (D) Wave IV, (E) Wave V, (F) Wave VI.

ABR WAVES

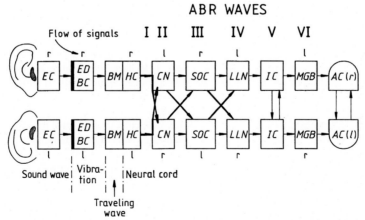

FIG. 2.14 Schematic illustration of the flow of signals in auditory pathways. The source location of each wave (ABR) was previously investigated for both animal and human subjects. *AC*, Auditory cortex of the right and left hemispheres; *BM* and *HC*, basilar membrane and hair cell; *CN*, cochlear nucleus; *EC*, external canal; *ED* and *BC*, eardrum and bone chain; *SOC*, superior olivary complex; *LLN*, lateral lemniscus nucleus; *IC*, inferior colliculus; *MGB*, medial denticulate body. (Jewett, D.L. (1970). Volume-conducted potentials in response to auditory stimuli as detected by averaging in the cat. *Electroenceph. Clin. Neurophysiol.*, 28, 609–618; Lev, A., and Sohmer, H. (1972). Sources of averaged neural responses recorded in animal and human subjects during cochlear audiometry (Electro-cochleogram). *Arch. Klin. Exp. Ohr., Nas.- u. Kehlk. Heilk.*, 201, 79–90 and Buchwald, J.A.S. and Huang, C.M. (1975). Far-field acoustic response: origins in the cat. *Science*, 189, 382–384.)

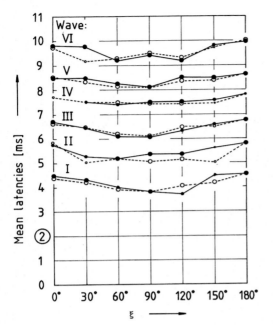

FIG. 2.15 Averaged latencies of auditory brainstem responses (ABRs) for four subjects, waves I–VI. Sizes of the *circles* indicate the number of available data from four subjects. The latency of 2 ms indicated by ② corresponds to the distance between the loudspeaker and the center of the head. (●), Left ABRs; (○), right ABRs.

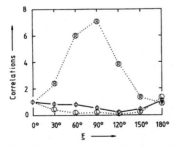

FIG. 2.16 Correlations of sound signals at the left- and right-ear entrances of a dummy head, which are normalized by the respective values at ξ = 0°. Ⓛ, Φll(0) measured at the left ear; Ⓡ, Φrr(0) measured at the right ear; Φ, maximum interaural cross-correlation, |Φlr(τ)|max, |τ| < 1 ms.

Although we cannot make a direct comparison between the results in Figs. 2.15 and 2.17, it is interesting to point out that the relative behavior of wave IVl is similar to Φrr(0) in Fig. 2.17, which was measured at the right ear entrance, r. Also, the relative behavior of

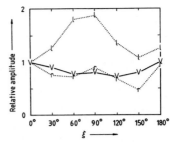

FIG. 2.17 Averaged amplitudes of waves IVl (symbol: l) and IVr (symbol: r), and averaged amplitudes of waves Vl and Vr (symbol: V) normalized to the amplitudes at the frontal incidence (four subjects).

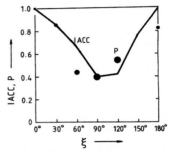

FIG. 2.18 Values of interaural cross-correlation function *(IACC)* and P calculated by Eqs. (4.15) and (2.2), respectively. A linear relationship between the IACC and the value *P* is obtained *(P < .005)*. Note that available datum at x = 150° was from a single subject.

wave IVr is similar to Φll(0) at the entrance to the left ear, l. In fact, amplitudes of wave IV (left and right) are proportional to Φlr(0), due to the interchange of signal flow. The behavior of wave V is similar to that of the maximum value, | Φlr(τ) |max, |τ| < 1 ms. Because correlations have the dimensions of the sound signal power, that is, the square of the ABR amplitude, the IACC defined by Eq. (4.15) has a corresponding neural measure of binaural correlation P

$$P = A_V^2 / [A_{IV,r} g A_{IV,l}] \qquad (2.2)$$

where A_V is the amplitude of wave V, which may be reflected by "maximum" neural activity (≈ | Φlr (τ) |max) at the inferior colliculus (see Fig. 2.16). Also, $A_{IV,r}$ and $A_{IV,l}$ are amplitudes of wave IV of the right and left, respectively. The results obtained by Eq. (2.2) are plotted in Fig. 2.18. It is clear that the relative behaviors of IACC and P are in good agreement, except for the value of P at ξ = 180° at which only a single datum for $A_{IV,r}$ was obtained, with only a single subject. The values

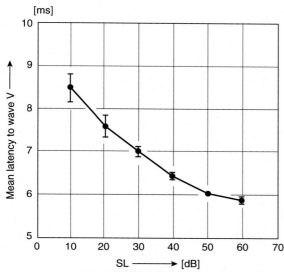

FIG. 2.19 Latency of ABR wave V as a function of SL in dB, i.e., sound detection threshold[14]. See FIG. 6.4 (A).

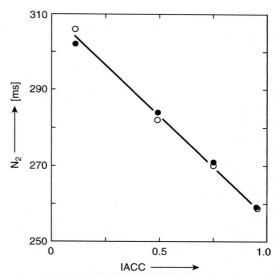

FIG. 2.20 Linear relationship between N2 and interaural cross-correlation function. See FIG. 6.4 (C).

exceeding unity are caused by error in the measurements. Obviously, a high correlation between the IACC and P values is achieved, that is, 0.92 ($P < .01$).

Remarks

The ABR amplitudes clearly differ according to the horizontal angle of the incidence of sound at the listener as shown in Fig. 2.13. In particular, we found that the amplitudes of waves IVl and IVr are nearly proportional to the sound pressures at the right and left ear entrances, respectively, when the amplitude is normalized to the amplitude in either the front ($\xi = 0°$) or back ($\xi = 180°$).

Concerning the behavior of the ABR amplitude on the left and right, the first interchange of the neural signal is considered to occur at the entrances of the cochlear nucleus, the second interchange may take place at the superior olivary complex, and the third may be at the lateral lemniscus nucleus as shown in Fig. 2.14. Thompson[13], who used neuroanatomical tract-tracing methods in guinea pigs, found four separate pathways connecting one cochlea with the other cochlea or with itself, all via brainstem neurons. This may relate to the first interchange at the entrance of the cochlear nucleus.

We discussed earlier that the "maximum" neural activity for wave V (inferior colliculus) in the auditory pathways corresponds to the IACC in the ABR amplitude, around 8.5 ms after the sound signal was supplied to the loudspeakers (68 ± 1 cm). However, the latency

of wave V decreases with increasing sensation level as shown in Fig. 2.19.[14] This implies binaural summation of the sound energy or the sound pressure level, which may be reflected in both $\Phi_{ll}(0)$ and $\Phi_{rr}(0)$ corresponding to Eq. (4.12). We will further discuss how the sound level response indicates right hemisphere dominance. Also, the relative latency at wave III corresponds to the interaural time difference (see Fig. 2.16).

We will also discuss remarkable evidence supporting a linear relationship between the IACC and the N2-latency observed in the slow vertex response (SVR) over both cerebral hemispheres ($P < .025$), as shown in Fig. 2.20. The subjective preference and subjective diffuseness judgments of the sound field described previously in relation to the IACC are thus well based on the neural activity of the auditory-brain system.

CENTRAL AUDITORY SIGNAL PROCESSING MODEL

Based on the above-mentioned physical system and physiological responses, a central auditory signal processing model was first formed for the temporal and spatial factors,[8] and was reconfirmed by neural evidences from a number of investigations.[15] The model consists of the autocorrelation mechanisms, the interaural cross-correlation mechanism between the two auditory pathways, and the specialization of human cerebral hemispheres for temporal and spatial factors

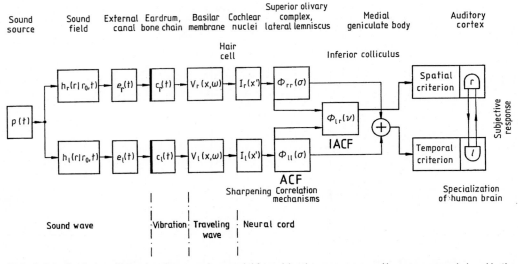

FIG. 2.21 Central auditory signal processing model for subjective responses. p(t): source sound signal in the time domain; hl,r(r/r0,t): head-related impulse responses from a source position of r0 to the left and right-ear entrances of a listener at r; el,r(t): impulse responses of left and right external canals, from the left- and right-ear entrances to the left and right eardrums; cl,r(t): impulse responses for vibration of left and right bone chains, from the eardrums to oval windows, including transformation factors into vibration motion at the eardrums: Vl,r(x,ω): travelling wave forms on the basilar membranes, where x is the position along the left and right basilar membrane measured from the oval window; Il,r(x'): sharpening in the cochlear nuclei corresponding to roughly the power spectra of input sound, that is, responses of a single pure tone ω tend to approach a limited region of nuclei. These neural activities may be enough to convert into activities similar to the autocorrelation function (ACF); Φll(σ) and Φrr(σ): ACF mechanisms in the left and right auditory pathways, respectively. Symbol ⊕ signifies that signals are combined; Φlr(ν): interaural cross-correlation function (IACF) mechanism; r and l: specialization for temporal and spatial factors of the left and right human cerebral hemispheres, respectively. Temporal sensations (see Chapter 7 of section "Temporal Primary Percepts in Relation to Temporal Factors of a Sound Signal") and spatial sensations (see Chapter 7 of section "Spatial Percepts in Relation to Spatial Factors of the Sound Field") may be processed in the left and right hemispheres according to the temporal factors extracted from the ACF and the spatial factors extracted from the IACF, respectively. The overall subjective responses, for example, subjective preference and annoyance, may be processed in both hemispheres in relation to the temporal and spatial factors (Ando, Y. (2002). Correlation factors describing primary and spatial sensations of sound fields. *Journal of Sound and Vibration*, 258, 405–417). (From Ando Y. *Concert Hall Acoustics*. Heidelberg: Springer-Verlag; 1985.)

of the sound field. In addition, according to the relationship of subjective preference and physiological phenomena in changes with variation to the temporal and spatial factors, a model may be proposed as shown in Fig. 2.21. In this figure, a sound source p(t) is located at r_0 in a three-dimensional space and a listener is sitting at r, which is defined by the location of the center of the head, $h_{l,r}(r| r_0,t)$ being the impulse responses between r_0 and the left and right ear-canal entrances. The impulse responses of the external ear canal and the bone chain are $e_{l,r}(t)$ and $c_{l,r}(t)$, respectively. The velocity of the basilar membrane is expressed by $V_{l,r}(x, \omega)$, with x being the position along the membrane.

The action potentials from the hair cells are conducted and transmitted to the cochlear nuclei, the superior olivary complex including the medial superior olive, the lateral superior olive, and the trapezoid body, and to the higher level of the two cerebral hemispheres. The input power density spectrum of the cochlea I(x') can be roughly mapped at a certain nerve position x',[16,17] as a temporal activity. Amplitudes of waves (I–IV) of the ABR reflect the sound pressure levels at both ears as a function of the horizontal angle of incidence to a listener (see Chapter 4 of section "Sound Transmission from a Point Source to Ear Entrances in an Enclosure"). Such neural activities, in turn, include

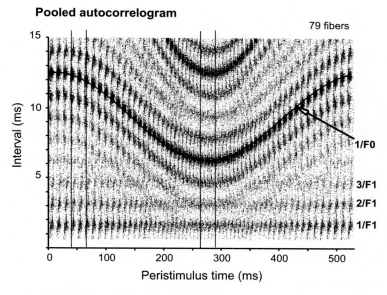

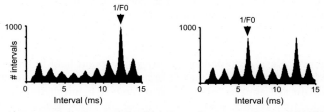

FIG. 2.22 Autocorrelograms and autocorrelation histograms in response to a variable F0 from 80 to 160 Hz of a single-formant vowel. (A) Pooled autocorrelograms for the vowel. (B) and (C) Smoothed autocorrelograms at two different regions. (From Cariani PA, Delgutte B. Neural correlates of the pitch of complex tones. II. Pitch shift, pitch ambiguity, phase-invariance, pitch circularity, and the dominance region for pitch. *J Neurophysiol.* 1996;76:1717−1734.)

sufficient information to attain the ACF, probably at the lateral lemniscus as indicated by $\Phi_{ll}(\sigma)$ and $\Phi_{rr}(\sigma)$. In fact, the time domain analysis of neural firing rates from the auditory nerve of cats reveals a pattern of ACF rather than the frequency domain analysis.[18] Pooled interspike interval distributions resemble the short time or the running ACF for the low-frequency component as shown in Fig. 2.22. It traces a change of the missing fundamental or pitch as a function of time. In addition, the pooled interval distributions for sound stimuli consisting of the high-frequency component resemble the envelope of the running ACF.[19] From the viewpoint of the missing fundamental or pitch of the complex tone judged by humans, the running ACF must be processed in the frequency components below

about 5 kHz.[20] The missing fundamental or pitch may be perceived below about 1.2 kHz (see Chapter 7 of section "Frequency Limits of the ACF Model for Pitch Percept"), which may cover most of musical signals. A tentative model of the running ACF processor is illustrated in Fig. 2.23. The output of the ACF processor may be dominantly connected with the left cerebral hemisphere.

We also discussed that specific neural activity (wave V together with waves IV$_l$ and IV$_r$) may correspond to the IACC as shown in Fig. 2.18. The interaural cross-correlation mechanism may thus exist at the inferior colliculus. We concluded that the output signal of the interaural cross-correlation mechanism including the IACC may be dominantly connected to the right

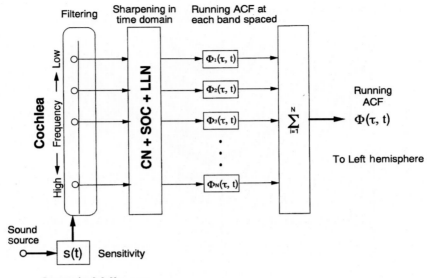

Filtering   Sharpening in   Running ACF at
             time domain   each band spaced

Running
ACF

$$\Phi(\tau, t)$$

To Left hemisphere

Outer/middle ear

FIG. 2.23 A tentative two-dimensional model of the autocorrelation function *(ACF)* processor in the auditory pathways. Each single neuron can act as a filter for its characteristic frequency. It is worth noticing that there are no scientific evidences for the fixed range of frequency band like the 1/3 octave or 1/1 octave filters, because cut-off frequencies might be arbitrary.

hemisphere. Also, the sound pressure level expressed by a geometrical average of the ACFs for the two ears at the origin of time ($\sigma = 0$), which in fact appears in the latency at the inferior colliculus, may be processed in the right hemisphere. Such a neural process has developed, realizing minimal effort and maximal efficiency, so that only critical information extracted from the ACF and IACF is transmitted into the left and right hemispheres, respectively.

We have also concluded that listening level (LL) and IACC are associated with the right cerebral hemisphere, and the temporal factors, Δt_1 and T_{sub}, of the sound field in a room are associated with the left (Table 6.2). The specialization of the human cerebral hemispheres may relate to the highly independent contribution between the spatial and temporal factors on any subjective attribute. For example, "cocktail party effects" might well be explained by such specialization of the human brain, because speech is processed in the left hemisphere, and independently, spatial information is mainly processed in the right hemisphere (see Chapter 6 of section "Specialization of Cerebral Hemispheres for Temporal and Spatial Factors").

Based on the model, we are able to describe temporal and spatial percepts, that is, subjective attributes of any given sound field in terms of processes occurring in the auditory pathways and the specialization of the two cerebral hemispheres, as well as subjective attributes of any visual field.[15]

REFERENCES

1. Mehrgardt S, Mellert V. Transformation characteristics of the external human ear. *J Acoust Soc Am.* 1977;61: 1567−1576.
2. Wiener FM, Ross DA. The pressure distribution in the auditory canal in a progressive sound field. *J Acoust Soc Am.* 1946;18:401−408.
3. Shaw EAG. Transformation of sound pressure level from the free field to the eardrum in the horizontal plane. *J Acoust Soc Am.* 1975;56:1848−1861.
4. Bekesy G. *Experiments in Hearing (Wever E, transl. and ed.).* New York: McGraw-Hill; 1960.
5. Tonndorf J, Khanna SM. Tympanic-membrane vibrations in human cadaver ears studied by time-averaged holography. *J Acoust Soc Am.* 1972;52:1221−1233.
6. Onchi Y. Mechanism of the middle ear. *J Acoust Soc Am.* 1961;33:794−805.
7. Rubinstrin M, Feldman B, Fischler H, Frei EH. Measurement of stapedial-footplate displacements during transmission of sound through the middle ear. *J Acoust Soc Am.* 1966;40:1420−1426.

8. Ando Y. *Concert Hall Acoustics*. Heidelberg: Springer-Verlag; 1985.

9. Puria S, Rosowski JJ, Peake WT. Middle-ear pressure gain in humans: preliminary results. In: Duifhuis H, Horst JW, Dijk P, Netten SM, eds. *Proceedings of the International Symposium on Biophysics of Hair Cell Sensory Systems*. World Scientific Publishing Co. 1993:345−351.

10. Zwislocki JJ. The acoustic middle ear function. In: Feldman AS, Wilber LA, eds. *Acoustic impedance and admittance—The measurement of middle ear function*. Baltimore: Williams and Wilkins Co. 1976; Chapter 4.

11. Berger EH. Re-examination of the low-frequency (50-1000 Hz) normal threshold of hearing in free and diffuse sound fields. *J Acoust Soc Am*. 1981;70:1635−1645.

12. Ando Y, Yamamoto K, Nagamastu H, Kang SH. Auditory brainstem response (ABR) in relation to the horizontal angle of sound incidence. *Acoustic Letters*. 1991;15:57−64.

13. Thompson AM, Thompson GC. Neural connections identified with PHA-L anterograde and HRP retrograde tract-tracing techniques. *J Neurosci Meth*. 1988;25:13−17.

14. Hecox K, Galambos R. Brain stem auditory evoked responses in human infants and adults. *Arch Otolaryngol*. 1974;99:30−33.

15. Ando Y, Cariani P, Guest, eds. *Auditory and Visual Sensations*. New York: Springer-Verlag; 2009.

16. Katsuki Y, Sumi T, Uchiyama H, Watanabe T. Electric responses of auditory neurons in cat to sound stimulation. *J Neurophysiol*. 1958;21:569−588.

17. Kiang NY-S. *Discharge Pattern of Single Fibers in the Cat's Auditory Nerve*. Cambridge, MA: MIT Press; 1965.

18. Secker-Walker HE, Searle CL. Time domain analysis of auditorynervefiber firing rates. *J Acoust Soc Am*. 1990;88:1427−1436.

19. Cariani PA, Delgutte B. Neural correlates of the pitch of complex tones. II. Pitch shift, pitch ambiguity, phase-invariance, pitch circularity, and the dominance region for pitch. *J Neurophysiol*. 1996;76:1717−1734.

20. Inoue M, Ando Y, Taguti T. The frequency range applicable to pitch identification based upon the auto-correlation function model. *J Sound Vibr*. 2001;241:105−116.

CHAPTER 3

Analysis of Sound Signals

ANALYSIS OF A SOURCE SIGNAL

Autocorrelation Function of a Sound Signal

The most promising signal in the auditory system to process after the peripheral power spectrum process is the autocorrelation function (ACF) of sound signals, which is defined by

$$\Phi_p(\tau) = \lim_{T \to \infty} \frac{1}{2T} \int_{-T}^{+T} p'(t)p'(t+\tau)dt \qquad (3.1)$$

where $p'(t) = p(t) * s(t)$, $s(t)$ is the sensitivity of the ear. For convenience, $s(t)$ can be chosen as the impulse response of an A-weighted network. It is worth noting that the physical system between the ear's entrance and the oval window has almost the same characteristics as the ear's sensitivity (see Chapter 2: Physical Systems of the Human Ear).[1,2]

It is well known that the ACF can also be obtained from the power density spectrum, so that

$$\Phi_p(\tau) = \int_{-\infty}^{+\infty} P_d(\omega)e^{j\omega t}dt \qquad (3.2)$$

$$P_d(\omega) = \int_{-\infty}^{+\infty} \Phi_p(\tau)e^{-j\omega t}d\tau \qquad (3.3)$$

Thus the ACF and the power density spectrum contain the same mathematical information. The normalized ACF is defined by

$$\phi_p(\tau) = \Phi_p(\tau)/\Phi_p(0) \qquad (3.4)$$

Five significant factors can be directly extracted from the short-time (2T) moving or the running ACF (Figs. 3.1−3.3), which cannot be obtained directly from the power spectrum:

1. Energy represented at the origin of delay, $\Phi_p(0)$;
2. As shown in Fig. 3.4, the width of amplitude $\phi(\tau)$, around the origin of the delay time, defined at a value of 0.5, is $W_{\phi(0)}$; it is noteworthy that $\phi(\tau)$ is an even function.

3. Two factors from the fine structure of ACF, including peaks and delays. For instance, τ_1 and ϕ_1 are the delay time and the amplitude of the first peak of the ACF, τ_n and ϕ_n are the delay time and the amplitude of the n-th peak, respectively. Usually there are certain correlations between τ_1 and τ_{n+1}, and between ϕ_1 and ϕ_{n+1}; thus two significant factors are τ_1 and ϕ_1;

4. The effective duration of the envelope of the normalized ACF, τ_e, is defined by the 10th-percentile delay (-10 dB) and represents a repetitive feature or reverberation containing the sound source itself (Fig. 3.5).

To exemplify two extremes, the ACF of pure tones for any phase is a cosine function with its period, τ_e, being infinite, whereas the ACF of white noise is $\delta(\tau)$, the Dirac delta function, and τ_e is zero. The ACF of complex tones is a linear superposition of cosines corresponding to the frequency components weighted by their amplitude, Thus τ_e is infinite and τ_1 corresponds to the pitch, as discussed in Chapter 7: Temporal Primary Percepts in Relation to Temporal Factors of a Sound Signal. The value of τ_e of general sound signals is between zero and infinity.

Running Autocorrelation Function of a Source Signal

A certain degree of coherence exists in the time sequence of the source signal, which may describe primary percepts of sound by the 2T moving ACF.

The 2T moving ACF as a function of time, τ, is calculated by[3]

$$\phi_p(\tau) = \phi_p(\tau; t, T) = \frac{\Phi_p(\tau; t, T)}{[\Phi_p(0; t, T)\Phi_p(0; \tau+t, T)]^{1/2}} \qquad (3.5)$$

where

$$\Phi_p(\tau; t, T) = \frac{1}{2T} \int_{t-T}^{t+T} p'(s)p'(s+\tau)ds \qquad (3.6)$$

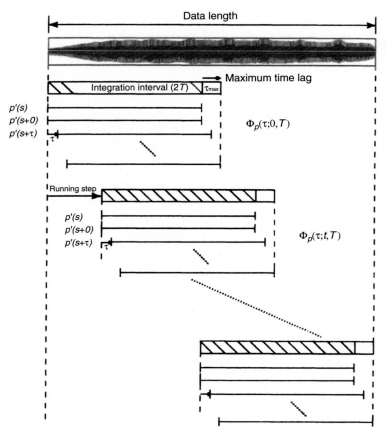

FIG. 3.1 Direct method of analyzing the running autocorrelation function (ACF) in the time domain.

The normalized ACF satisfies the condition that $\phi_p(0) = 1$.

In order to demonstrate a procedure for obtaining the effective duration of the analyzed 2T ACF, Fig. 3.5 shows the absolute value in logarithmic form as a function of the delay time. The envelope decay of the initial and important part of the ACF may be fitted by a straight line in most cases. The effective duration of the ACF is defined by the delay τ_e, at which the envelope of the ACF becomes -10 dB (or 0.1, the 10th-percentile delay). This can be easily obtained, for example, by the decay rate extrapolated in the range from 0 dB at the origin to -5 dB.

Table 3.1 lists the temporal monaural factors that are extracted from the running ACF.

VOCAL SOURCE SIGNALS

In order to demonstrate a procedure for extracting the effective duration from the analyzed running ACF, Fig. 3.5 shows the absolute value in logarithmic form as a function of the delay time.[4] The envelope decay of initial and important parts of the running ACF may be fitted by a straight line in the range of 5 dB for most cases, as shown in the figure. In case the 5-dB range is unavailable, the value of τ_e is obtained

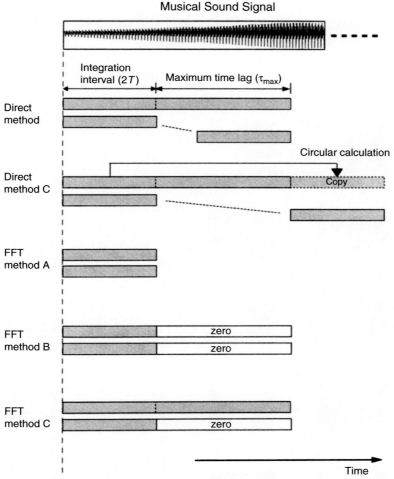

FIG. 3.2 Five methods for analyzing the ACF. (Kato K, Fujii K, Hirawa T, Kawai K, Yano T, Ando Y. Investigation of the relation between minimum effective duration of running autocorrelation function and operatic singing with different interpretation styles. *Acta Acust United Ac.* 2007;93:421−434.)[5]

by the initial 5-ms delay interval as far as speech signals are concerned.

Examples of analyzed τ_e values for vowel signals sung by a soprano are demonstrated in Fig. 3.6. Although τ_e values are subject to variation according to 2T, the most important minimum value as well as local minima are independent in a certain range of 2T for vocal signals.

RECOMMENDED AUDITORY-TEMPORAL WINDOW

In analyzing the running ACF, the "auditory-temporal window" 2T in Eqs. (3.5) and (3.6) must be carefully determined. The initial part of ACF within the effective duration, τ_e, of the ACF contains important information regarding the signal. In order to determine the auditory-temporal window, successive loudness judgments in

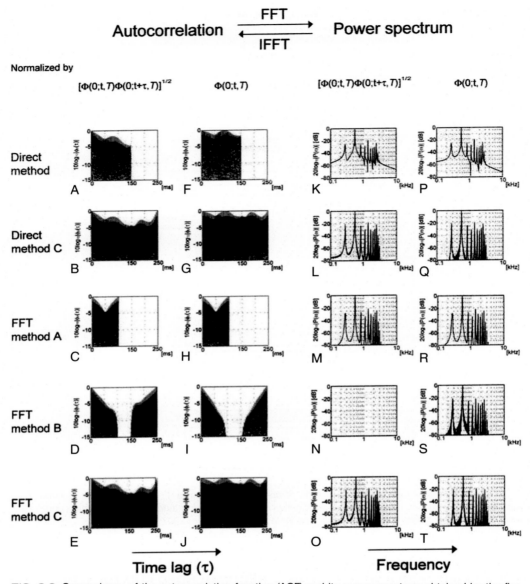

FIG. 3.3 Comparisons of the autocorrelation function (ACF) and its power spectrum obtained by the five methods shown in Fig. 3.2. Fast Fourier Transformation (FFT) methods (A) and (B) cannot obtain the right ACF. The FFT method (C) is recommended for analyzing the ACF up to the maximum delay time, τ_{max}.

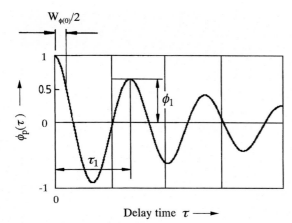

FIG. 3.4 Significant factors $W_{\phi(0)}$, τ_1, and ϕ_1 extracted from the initial part of the running autocorrelation function (ACF).

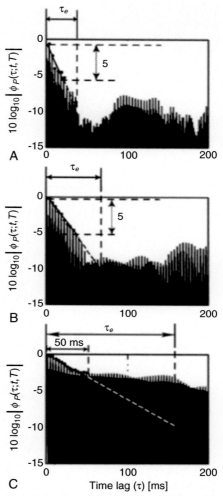

FIG. 3.5 Examples of the autocorrelation function (ACF) in logarithmic form and the extracted τ_e value, defined as the delay at -10 dB. (A) and (B) Extrapolations of the ACF at -10 dB due to envelope decay to -5 dB. (C) If envelope decay to -5 dB is unavailable, then extrapolation of the envelope decay for 50 ms can be made.

pursuit of the running listening level (LL) have been conducted. Results are shown in Fig. 3.7. A recommended signal duration $(2T)_{recom.}$ to be analyzed is approximately given by

$$(2T)_{recom.} : \sim 30(\tau_e)_{min} \qquad (3.7)$$

where $(\tau_e)_{min}$ is the minimum value of τ_e obtained by analyzing the ACF. This signifies an adaptive temporal window depending on the temporal activity of the sound signal in the auditory system. For example, the temporal windows may differ widely according to music pieces ($[2T]_r \sim 0.5 - 5$ s) and vowels ($[2T]_r \sim 40 - 100$ ms) and consonants ($[2T]_r \sim 5 - 10$ ms) in a continuous speech signal. This indicates that the brain may be more relaxed when listening to slow music than when listening to speech. In other words, more concentration is invested in listening to speech than in listening to music. Note that the running step (R_s), which signifies a degree of overlap of signal to be analyzed, is not critical. It may be selected as $K_2(2T)_{recom.}$, K_2 being chosen, say, in the range of 1/4 to 1/2.

TABLE 3.1
Temporal monaural factors directly extracted from the running ACF

Temporal factors	Symbols	Primary temporal percepts
(1) Sound energy	$\Phi_p(0)$	Loudness
(2) Tilt of spectrum	$W_{\phi(0)}$	Timbre
(3) Delay time of the first peak	τ_1	Pitch
(4) Amplitude of the first peak	$\phi_1{}^1$	Pitch strength
(5) Effective duration of ACF envelope	τ_e	$(\tau_e)_{min}$: Segmentations of V & CV; preferred conditions of the temporal factors of sound fields

[1] Another significant factor is $\Delta\phi_1/\Delta t$, i.e., the speech of development of the pitch, which depends on speed of pronunciation.

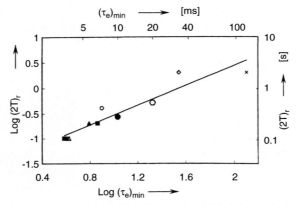

FIG. 3.6 Examples of the measured τ_e value extracted from the running ACF of 20 vowels sung by a soprano singer with four different pitches obtained for three different signal durations. Thin curve, 2T = 100 ms; dotted curve, 2T = 200 ms; thick curve, 2T = 500 ms. (Kato K, Fujii K, Hirawa T, Kawai K, Yano T, Ando Y. Investigation of the relation between minimum effective duration of running autocorrelation function and operatic singing with different interpretation styles. *Acta Acust United Ac.* 2007; 93:421−434.)

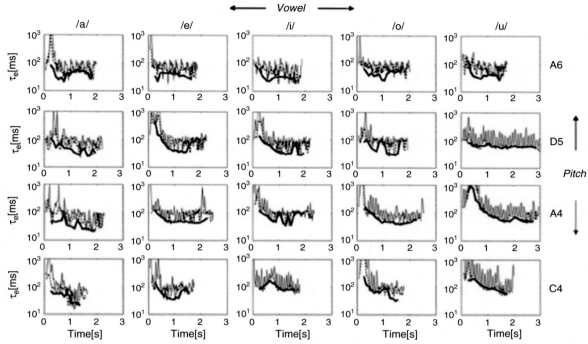

FIG. 3.7 Recommended signal duration to be analyzed for obtaining the autocorrelation function (ACF). (Mouri K, Akiyama K, Ando Y. Preliminary study on recommended time duration of source signals to be analyzed, in relation to its effective duration of autocorrelation function. *J Sound Vib.* 2001;241:87—95.)

REFERENCES

1. Ando Y. *Concert Hall Acoustics.* Heidelberg: Springer-Verlag; 1985.
2. Ando Y. *Architectural Acoustics, Blending Sound Sources, Sound Fields, and Listeners.* New York: AIP Press/Springer-Verlag; 1998.
3. Taguti T, Ando Y. Characteristics of the short-term autocorrelation function of sound signals in piano performances. In: Ando Y, Noson D, eds. *Music and Concert Hall Acoustics, Conference Proceedings of MCHA 1995.* London: Academic Press; 1997. Chapter 23.

4. Kato K, Fujii K, Hirawa T, Kawai K, Yano T, Ando Y. Investigation of the relation between minimum effective duration of running autocorrelation function and operatic singing with different interpretation styles. *Acta Acust United Ac.* 2007;93:421—434.
5. Mouri K, Akiyama K, Ando Y. Preliminary study on recommended time duration of source signals to be analyzed, in relation to its effective duration of autocorrelation function. *J Sound Vib.* 2001;241:87—95.

CHAPTER 4

Formulation and Simulation of Sound Fields in an Enclosure

SOUND TRANSMISSION FROM A POINT SOURCE TO EAR ENTRANCES IN AN ENCLOSURE

Let us consider sound transmission from a source point in a free field to binaural ear canal entrances. Let $p(t)$ be the source signal as a function of time t, and $g_l(t)$ and $g_r(t)$ be impulse responses between the source point r_0 and the binaural entrances of a listener. Then the sound signals arriving at the entrances are expressed by[1]

$$f_l(t) = p(t) * g_l(t)$$
$$f_r(t) = p(t) * g_r(t)$$
$$(4.1)$$

where the asterisk denotes convolution. The impulse responses $g_{l,r}(t)$ consist of the direct sound and reflections $w_n(t - \Delta t_n)$ from the walls in the room as well as the head-related impulse responses $h_{nl,r}(t)$, so that

$$g_{l,r}(t) = \sum_{n=0}^{\infty} A_n w_n(t - \Delta t_n) * h_{nl,r}(t) \qquad (4.2)$$

where n denotes the number of reflections with horizontal angle ξ_n and elevation η_n, $n = 0$ signifies the direct sound ($\xi_0 = 0, y_0 = 0$):

$$A_0 w_0(t - \Delta t_0) = \delta(t), \Delta t_0 = 0, A_0 = 1$$

where $\delta(t)$ is the Dirac delta function, A_n is the pressure amplitude of the n-th reflection $n > 0$ in reference to that of the direct sound A_0, $w_n(t)$ is the impulse response of the walls for each path of reflection arriving at the listener, Δt_n is the delay time of reflection relative to that of the direct sound, and $h_{nl,r}(t)$ are impulse responses for diffraction of the head and pinnae for the single sound direction of n. Therefore Eq. (4.1) becomes

$$f_{l,r}(t) = \sum_{n=0}^{\infty} p(t) * A_n w_n(t - \Delta t_n) * h_{nl,r}(t) \qquad (4.3)$$

When the source has directivity characteristics, then $p(t)$ is replaced by $p_n(t)$.

ORTHOGONAL FACTORS OF THE SOUND FIELD

According to sound transmission from a point source to ear entrances in the sound field of an enclosure as mentioned in the previous section, we can determine the orthogonal factors consisting of temporal and spatial factors of the sound field.

Temporal Factors of the Sound Field

The temporal factor is extracted from the set of impulse responses of the reflecting walls, $A_n(t - \Delta t_n)$ of the sound field. The amplitudes of the reflections relative to that of the direct sound A_0; $A_1, A_2...$ are determined by the pressure decay due to the paths d_n, such that

$$A_n = d_0/d_n \qquad (4.4)$$

where d_0 is the distance between the source point and the center of the listener's head. The impulse responses of reflections to the listener are $w_n(t - \Delta t_n)$ with the delay times of $\Delta t_1, \Delta t_2, ...$ relative to that of the direct sound, which are given by

$$\Delta t_n = (d_n - d_0)/c \qquad (4.5)$$

where c is the velocity of sound [m/s].

These parameters are not physically independent; in fact the values of A_n are directly related to Δt_n in a relationship given by

$$\Delta t_n = d_0(1/A_n - 1)/c \qquad (4.6)$$

In addition, the initial time daelay gap between the direct sound and the first reflection Δt_1 is statistically related to $\Delta t_2, \Delta t_3, ...,$ which depend on the dimensions and the shape of the room. In fact, the echo density is proportional to the square of the time delay.[2] Thus the initial time delay gap Δt_1 is regarded as a representation of both sets of Δt_n and A_n ($n = 1, 2,...$).

Another parameter is the set of the impulse responses of the n-th reflection, $w_n(t)$ being expressed by

$$w_n(t) = w_n(t)^{(1)} * w_n(t)^{(2)} * \cdots * w_n(t)^{(i)} \qquad (4.7)$$

where $w_n(t)^{(1)}$ is the impulse response of the i-th wall existing in the path of the n-th reflection from the source to the listener.

Such a set of impulse responses $w_n(t)^{(1)}$ may be represented by a statistical decay rate, namely the subsequent reverberation time, T_{sub}, because $w_n(t)^{(1)}$ includes the absorption coefficient $\alpha_n(\omega)^{(i)}$ as a function of frequency. This coefficient is given by

$$\alpha_n(\omega)^{(i)} = 1 - \left|W_n(\omega)^{(i)}\right|^2 \qquad (4.8)$$

It is worth noting that as far as a single reflection is concerned, the most preferred condition of $w_n(t)^{(1)}$ has been examined, with the perfect reflection given by $\delta(t)$ the most preferred condition.[1]

According to the Sabine formula (1900),[3] the subsequent reverberation time is approximately calculated by

$$T_{sub} \approx \frac{KV}{\overline{\alpha}S} \qquad (4.9)$$

where K is a constant (approximately 0.162), V is the volume of the room, S is the total surface, and $\overline{\alpha}$ is the average absorption coefficient of the walls, and $\overline{\alpha}S$ is given by the summation of the absorption of each surface i, so that

$$\overline{\alpha}S = \sum_i \alpha(\omega)^{(i)}S^{(i)} \qquad (4.10)$$

So far, we figured out the significant temporal factors of the sound field:
1. The initial delay time of the first reflection, Δt_1 given by Eq. (4.6), $n = 1$.
2. The subsequent reverberation time, T_{sub} expressed by Eq. (4.9).

Spatial Factors of the Sound Field

Two sets of head-related impulse responses for two ears $h_{nl,r}(t)$ constitute the spatial factors. These two responses, $h_{nl}(t)$ and $h_{nr}(t)$, play an important role in sound localization and spatial impression or subjective diffuseness but are not mutually independent objective factors. Therefore, to represent the interdependence between two impulse responses, a single factor may be introduced, that is, the interaural cross-correlation function (IACF) between the sound signals at both ears $f_l(t)$ and $f_r(t)$, which is defined by

$$\Phi_{lr}(\tau) = \lim_{T \to \infty} \frac{1}{2T} \int_{-T}^{+T} f'_l(t)f'_r(t + \tau)dt, \quad |\tau| \leq 1 \text{ ms} \quad (4.11)$$

where $f'_l(t)$ and $f'_r(t)$ are obtained by signals $f_{l,r}(t)$ after passing through the A-weighted network, which corresponds to the ear's sensitivity, $s(t)$. We showed in

Chapter 2 of section "Physical Systems of the Human Ear" that ear sensitivity may be characterized by the physical ear system including the external and the middle ear.[1,4]

The normalized IACF is defined by,

$$\phi_{lr}(\tau) = \frac{\Phi_{lr}(\tau)}{\sqrt{\Phi_{ll}(0)\Phi_{rr}(0)}} \qquad (4.12)$$

where $\Phi_{ll}(0)$ and $\Phi_{rr}(0)$ are the ACFs at $\tau = 0$ for the left and right ear, respectively, or the sound energies arriving at both ears, and τ the interaural time delay possibly within plus and minus 1 ms. In addition, from the denominator of Eq. (4.12), we obtain the binaural listening level (LL) such that

$$LL = 10 \log \left[\Phi(0)/\Phi(0)_{reference}\right] \qquad (4.13)$$

where $\Phi(0) = [\Phi_{ll}(0) \Phi_{rr}(0)]^{1/2}$ is the geometrical mean of the sound energies arriving at the two ears and $\Phi(0)_{reference}$ is the reference sound energy.

If discrete reflections arrive after the direct sound, then the normalized interaural cross-correlation (IACC) is expressed by

$$\phi_{lr}^{(N)}(\tau) = \frac{\sum\limits_{n=0}^{N} A^2 \phi_{lr}^{(n)}(\tau)}{\sqrt{\sum\limits_{n=0}^{N} A^2 \Phi_{ll}^{(n)}(0) \sum\limits_{n=0}^{N} A^2 \Phi_{rr}^{(n)}(0)}} \qquad (4.14)$$

where we put $w_n(t) = \delta(t)$ for the sake of convenience, and $\Phi_{lr}^{(n)}(\tau)$ is the IACC of the n-th reflection, $\Phi_{ll}(0)^{(n)}$ and $\Phi_{rr}(0)^{(n)}$ are the respective sound energies arriving at the two ears from the n-th reflection. The denominator of Eq. (4.14) corresponds to the geometric mean of the sound energies arriving at the two ears.

The magnitude of the IACC is defined by

$$IACC = |\phi_{lr}(\tau)|_{max} \qquad (4.15)$$

for the possible maximum interaural time delay,

$$|\tau| \leq 1 \text{ ms}$$

For several music motifs, the long-time IACF ($2T = 35$ s) was measured for each single reflected sound direction arriving at a dummy head (Table D.1).[1] These data may be used for the calculation of the IACF by Eq. (4.14). For example, measured values of the IACF using music motifs A and B are shown in Fig. 4.1A and B.

The interaural delay time, at which the IACC is defined as shown in Fig. 4.2, is the τ_{IACC}. Thus both the IACC and τ_{IACC} may be obtained at the maximum value of IACF.

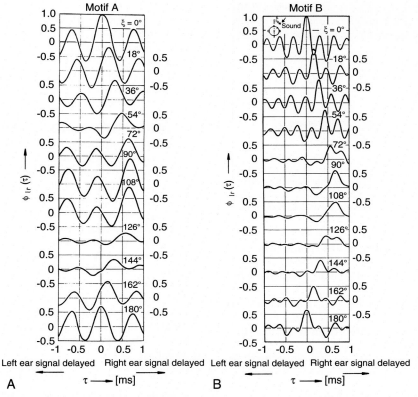

FIG. 4.1 Measured interaural cross-correlation function for sound without reflection as a parameter of horizontal angle of incidence ξ. (A) Music motif A. (B) Music motif B.

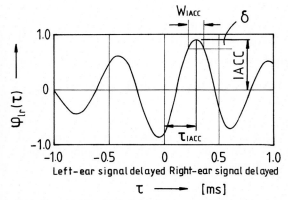

FIG. 4.2 Definition of the spatial factors, IACC, τ_{IACC} and W_{IACC}, extracted from the interaural cross-correlation function.

For a single source signal arriving from the horizontal angle ξ defined by τ_ξ, the interaural time delay corresponds to τ_{IACC}. When we observe $\tau_{IACC} = 0$ in an enclosure, then usually a frontal sound image and a well-balanced sound field to the left and right are perceived (the preferred condition).

The width of the IACF, defined by the interval of delay time at a value of δ below the IACC, corresponding to the just noticeable difference (JND) of the IACC, is given by the W_{IACC} (see Fig. 4.2). A well-defined directional impression corresponding to the interaural time delay τ_{IACC} is perceived when listening to white noise with a sharp peak in the IACF with a small value of W_{IACC}. Thus the apparent source width (ASW) may be perceived as a directional range corresponding mainly to the W_{IACC}. On the other hand, when listening to a sound field with a low value for the IACC (< 0.15), a subjectively diffuse sound is perceived.[5] These four factors, LL, IACC, τ_{IACC}, and W_{IACC}, are independently related to spatial percepts such as subjective diffuseness and the ASW (see Chapter 7: MEG Responses to IACF Factors). These four factors, LL, IACC, τ_{IACC}, and W_{IACC}, are independently related to spatial percepts such as subjective diffuseness and the ASW (Chapter 7).[6,7]

TABLE 4.1
Spatial binaural factors extracted from the IACF

Spatial factors	Symbols	Primary spatial percepts
(1) Binaural sound energy	$[\Phi_{ll}(0)\Phi_{rr}(0)]^{1/2}$	Binaural loudness
(2) Magnitude of the IACF	IACC	Subjective diffuseness
(3) Width of IACC	W_{IACC}[1]	ASW
(4) Delay time of the peak	τ_{IACC}	Horizontal localization of the sound source

[1] It is worth noticing that $W_{IACC} \sim W_{\phi(0)}$ that is extracted from the ACF (Table 3).

Significant spatial factors of the sound field are extracted from the IACF, as listed in Table 4.1.

1. The binaural LL is obtained accurately as defined by Eq. (4.12).
2. The IACC is defined by Eq. (4.15) as defined in Fig. 4.2.
3. W_{IACC} is the width of IACC defined by $\delta = 0.1$ (approximately the JND of IACC).
4. The interaural delay time is τ_{IACC} at which IACC is obtained.

Auditory Time Window for IACF Processing

The previously listed spatial sensations may be judged immediately when we come into a sound field, because our binaural system is able to process the IACF in a short time window. This is quite difference from the adaptive temporal window for sound signals, which varies due to the effective duration of the ACF of the sound-source signal.

When a sound signal is moving in a horizontal direction on stage, we must select a suitable short "time window" 2T for analyzing the running IACF, which depends on the speed of the moving image in sound localization. The range of τ_{IACC} extracted from the IACF can describe the movement range of such an image. It is obvious that the range of τ_{IACC} cannot be obtained when the integration interval (2T) of the IACF is longer than the period of movement; on the other hand, the value of τ_{IACC} is too fluctuant to be determined when 2T is selected shorter than the possible maximum value of $\tau_{IACC} < 1$ ms. For a sound source moving sinusoidally in the horizontal plane with less than 0.2 Hz, 2T may be selected in a wide range from 30 to 1000 ms. In addition, when a sound source is moving in a range of 4.0 Hz, 2T = 30 to 100 ms is

acceptable.[8] For reliable results, to cover a wide range of movement velocities for horizontal localization, we recommend a temporal window for the IACF to be fixed at approximately 30 ms.

For a sound source fixed on stage, for example, the value of (2T) may be selected longer than 1.0 s for the measurement of spatial factors at each audience seat.

SIMULATION OF SOUND LOCALIZATION

The directional information in simulating the sound field in an opera house can be realized by means of spatial factors that are extracted from the IACF. Schroeder (1962)[9] first simulated sound localization in the horizontal plane by use of two loudspeaker reproduction systems. To make the perception correspond precisely to the actual direction of a sound source located at any position in a three-dimensional space, a general system considering the asymmetry of our head and pinnae[10] is described as follows.

Referring to the lower part of Fig. 4.3, let the pressure impulse response for the paths from the two loudspeakers L_1 and L_2 to the entrances of the left and right ear canals be $h_{l,r1}(t)$ and $h_{l,r2}(t)$, respectively. Then the pressures to be reproduced at the two entrances are expressed by

$$f_l(t) = x_1(t) * h_{l1}(t) + x_2(t) * h_{l2}(t)$$
$$f_r(t) = x_1(t) * h_{r1}(t) + x_2(t) * h_{r2}(t) \tag{4.16}$$

where x_1 and x_2 are the input signals supplied for the loudspeakers L_1 and L_2, respectively.

Fourier transforming both sides of Eq. (4.16) yields

$$F_l(\omega) = X_1(\omega)H_{l1}(\omega) + X_2(\omega)H_{l2}(\omega)$$
$$F_r(\omega) = X_1(\omega)H_{r1}(\omega) + X_2(\omega)H_{r2}(\omega). \tag{4.17}$$

Thus the input signals to the loudspeakers are given by

$$X_1(\omega) = [F_l(\omega) H_{r2}(\omega) - F_r(\omega)H_{l2}(\omega)]D(\omega)^{-1}$$
$$X_2(\omega) = [F_r(\omega) H_{l1}(\omega) - F_l(\omega)H_{r1}(\omega)]D(\omega)^{-1} \tag{4.18}$$

where $D(\omega) = H_{l1}(\omega)H_{r2}(\omega) - H_{r1}(\omega)H_{l2}(\omega)$.

Therefore the signals in the time domain to be fed into two loudspeakers are obtained by the inverse Fourier transform, such that

$$x_1(t) = [f_l(t)*h_{r2}(t) - f_r(t)*h_{l2}(t)] * d(t)$$
$$x_2(t) = [f_2(t)*h_{l1}(t) - f_l(t)*h_{r1}(t)] * d(t) \tag{4.19}$$

where $d(t)$ is the inverse Fourier transform of $D(w)^{-1}$. The necessary and sufficient condition for a unique solution is $D(\omega) \neq 0$, throughout the reproduced frequency range.

According to Eq. (4.19), it is easy to draw a block diagram of the reproduction filter as shown in Fig. 4.4

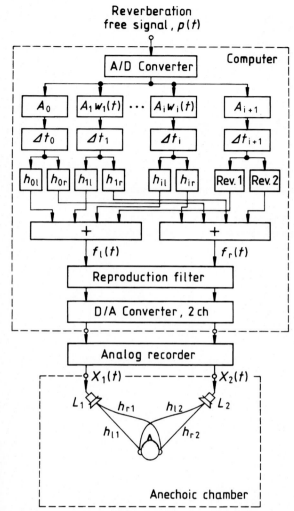

FIG. 4.3 A system of simulating the sound field with a direct sound and two early reflections, two incoherent reverberators in an enclosure and a reproduction system with two loudspeakers.

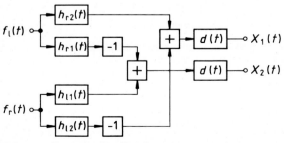

FIG. 4.4 Reproduction filter for two loudspeaker systems for the two ears.

to simulate the sound localization, and thus the sound field in a room may be simulated as shown in Fig. 4.3.

Let us consider the simplest sound field, a single sound source located at an arbitrary position in a free field. The sound pressure at the two ear entrances expressed by Eq. (4.3) may be reduced into a simple form, such that

$$f_{l,r}(t) = p(t) * h_{nl,r}(t) \qquad (4.20)$$

where $h_{nl,r}(t) \equiv h_{l,r}(\xi, \eta; t)$ are impulse responses between the source and the ear entrances. The head-related transfer functions (HRTFs) required for the filter shown in Figs. 4.3 and 4.4 are measured for each individual. In the experiment, two loudspeakers are located above the listener at angles $\xi = \pm 30$ degrees, $\eta = 90$ degrees, as shown in Fig. 4.5. In these conditions, the HRTF was fairly flat with no zeros and no significant dips for any participating subject.

This fact satisfies the condition for a unique solution as mentioned later in Eq. (4.19). Sound localization with an external sound image is created with a minimum resolution of 15 degrees, in the horizontal plane (ξ) and the median plane (η). Responses are shown in Fig. 4.6, as well as localization with real sound sources for three subjects with different-sized pinnae.[11] In this experiment, white noise (0.3–13.6 kHz) is presented as a source signal. By use of individual HRTF in the simulation, the accuracy of localization was almost the same order as for the real sound source. If we apply the HRTF from the other person, then the subject's accuracy in localization is generally decreased and, in some cases, localization is not possible.

Applying the reproduction system, the sound field in a real opera house with scattered and diffusing elements may be evaluated subjectively. When two loudspeakers are closely spaced, for example, $\xi = \pm 5$ degrees, $\eta = 0$ degrees, it is known as a "stereo-dipole" system.[12] A merit of the nonindividualized system is that it is uncritical in localization even in case of head movement during listening. In addition, it has been reported that by use of this system one can even distinguish an auditory distance from judgment of room size by mere perception.[13]

SIMULATION OF THE REVERBERANT SOUND FIELD

An example of a simulation system for the sound field in a room is shown in Fig. 4.3, based on Eq. (4.3). A reverberation-free vocal or orchestral music signal is applied to p(t). The program provides the amplitude and delay time of early reflections including directional information, and the subsequent reverberation. All data

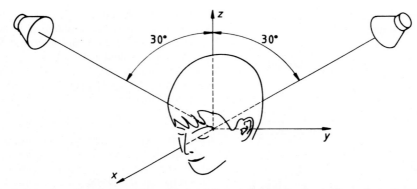

FIG. 4.5 Location of two loudspeakers for simulating sound localization in three-dimensional space.

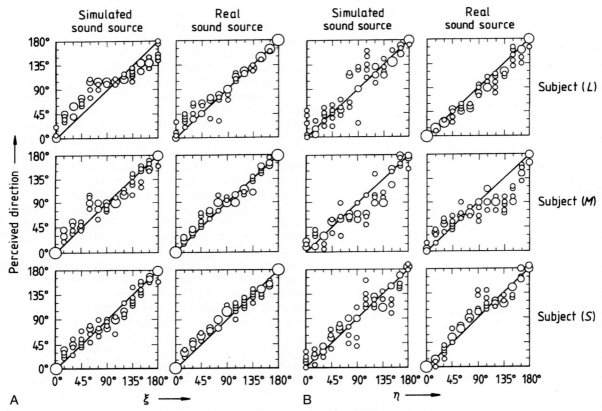

FIG. 4.6 Results of sound localization tests by three listeners with different sized pinnae for simulated sound source and real sound source. (A)Horizontal plane. (B) Median plane. (Morimoto M, Ando Y. On the simulation of sound localization. *JASJA.* 1980;1:167–174.)

are calculated relative to the direct sound (n = 0). As shown in the first column of the upper half part of Fig. 4.3, the direct sound is simulated by using only the HRTF to the two ears for the frontal direction, that is,

$$p(t) * h_{0l,r}(t) \qquad (4.21)$$

with $A_0 = 1$ and $\Delta t_0 = 0$. The second column simulates the first reflection (n = 1) for the two ears, which is given by

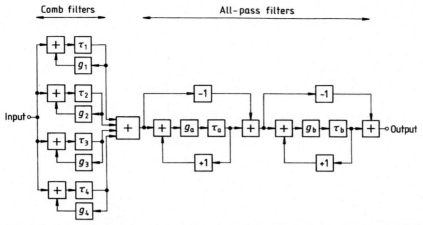

FIG. 4.7 Block diagram of a reverberator with four comb filters adjusting the subsequent reverberation time and two all-pass filters producing a proper density of reflections. (Schroeder MR. Natural sounding artificial reverberation. *J Audio Eng Soc.* 1962;10:219–223.)

$$p(t) * A_1 w_1(t - \Delta t_1) * h_{1,l,r}(t) \qquad (4.22)$$

Similarly, two early reflections are simulated, which can usually be distinguished in the impulse response measured in rooms. After these early reflections, two incoherent reverberation signals are applied.

A block diagram of a reverberator is shown in Fig. 4.7.[9] These sound signals simulated for the left and right ears are added respectively and fed into the reproduction filter as shown in Fig. 4.4. The reverberator consists of comb filters and all-pass filters. The impulse response of one of the comb filters with delay, τ, and gain, g, as shown in Fig. 4.8 is expressed by

$$h(t) = \delta(t - \tau) + g\delta(t - 2\tau) + g^2\delta(t - 3\tau) + \cdots \quad (4.23)$$

so that the reflections decrease exponentially. The Fourier transform of Eq. (4.23) gives corresponding frequency characteristics, such as

$$H(\omega) = e^{-j\omega\tau} + ge^{-j2\omega\tau} + g^2 e^{-j3\omega\tau} + \ldots = e^{-j\omega\tau}/(1 - g\,e^{-j\omega\tau})$$
$$(4.24)$$

The absolute value of $H(\omega)$, which is given by

$$|H(\omega)| = 1/[1 + g^2 - 2g\cos\omega\tau]^{1/2} \qquad (4.25)$$

is shown in the lower part of Fig. 4.8. The amplitude as a function of frequency presents a comb with periodic structure. For $\omega = 2n\pi/\tau$, n = 0, 1, 2... and g > 0, it has the maxima, so that

$$|H(\omega)|_{min} = 1/(1 - g) \qquad (4.26)$$

And for $\omega = (2n + 1)\,\pi/\tau$, n = 0, 1, 2... the minima

$$|H(\omega)|_{max} = 1/(1 + g) \qquad (4.27)$$

Thus the ratio between the maxima and minima yields

$$|H(\omega)|_{max}/|H(\omega)|_{min} = (1 + g)/(1 - g) \qquad (4.28)$$

For example, if g = 0.85, then the ratio is 12.3 or 22 dB. This produces a "colored" and "fluttered" quality. By

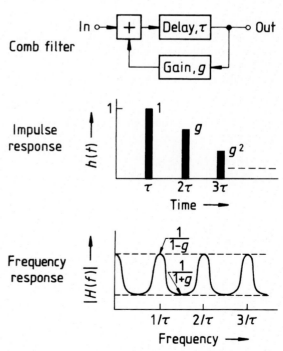

FIG. 4.8 Comb filter, the impulse response, and the frequency characteristics. (Schroeder MR. Natural sounding artificial reverberation. *J Audio Eng Soc.* 1962;10:219-223.)

use of several different comb filters connected in parallel as shown in Fig. 4.7, highly irregular frequency response may help to avoid such an undesired phenomenon. Reverberation time is given by the loop gains g_1, $g_2, \ldots g_M$ and delays $\tau_1, \tau_2 \ldots \tau_M$ of the different comb filters. A sound level decay by $-20\log(g_m)$[dB] for every trip around the feedback loop τ_m gives

$$T_m = 60\tau_m/[-20\log(g_m)] = 3\tau_m/[-20\log(g_m)],$$
$$m = 1, 2, \ldots M$$

(4.29)

And the subsequent reverberation time yields

$$T_{sub} = [T_m]_{max}$$ (4.30)

Note that the reverberation time as a function of frequency can be realized by the impulse response $g_m(t)$ or its Fourier transform $G_m(\omega)$, which corresponds to the transfer function for reflection from the boundary wall.

To simulate a high density of reflections of order t^2, two all-pass filters are connected in series (see Fig. 4.7). The density of reflections at time, t, after the impulse excitation is given by

$$n_e(t) = \frac{1}{2} \sum_{m=1}^{M} \frac{1}{\tau_m} \frac{1}{\tau_a \tau_b} t^2$$ (4.31)

The delays τ_a and τ_b of the all-pass filters should be chosen as and τ_b much greater than τ_m, m = 1,2... M, so that they do not influence the reverberation time itself given by Eq. (4.29).

The impulse response of the all-pass filter is given by

$$h(t) = -g\delta(t) + (1 - g^2)[\delta(t - \tau) + g\delta(t - 2\tau) + \cdots]$$ (4.32)

Taking Eq. (4.24) into account, the Fourier transform of Eq. (4.32) yields

$$H(\omega) = -g + (1 - g)e^{-j\omega\tau}/(1 - ge^{-j\omega\tau})$$
$$= e^{-j\omega\tau}(1 - ge^{j\omega\tau})/(1 - ge^{-j\omega\tau})$$

Thus

$$|H(\omega)| = 1.0$$

for all frequencies.

In the reverberator shown in Fig. 4.7, if we set $g_a = g_b = 1/\sqrt{2}$ (≈ 0.7), then the all-pass filter is realized. It is worth noting that preferred spectra of the single reflection and of reverberation time are just "flat."

REFERENCES

1. Ando Y. *Concert Hall Acoustics*. Heidelberg: Springer-Verlag; 1985.
2. Kuttruff H. *Room Acoustics*. 3rd ed. London: Elsevier Applied Science; 1991.
3. Sabine WC. Reverberation. In: *The American Architect and the Engineering Record*, Collected papers on acoustics. Los Altos, California: Peninsula Publishing; 1900 [Chapter 1], Prefaced by Beranek, L.L. (1992).
4. Ando Y. *Architectural Acoustics, Blending Sound Sources, Sound Fields, and Listeners*. New York: AIP Press/Springer-Verlag; 1998.
5. Damaske P, Ando Y. Interaural cross-correlation for multichannel loudspeaker reproduction. *Acustica*. 1972;27: 232–238.
6. Ando Y, Sato S, Sakai H. Fundamental subjective attributes of sound fields based on the model of auditory-brain system. In: Sendra JJ, ed. *Computational Acoustics in Architecture*. Southampton: WIT Press; 1999.
7. Ando Y. Correlation factors describing primary and spatial sensations of sound fields. *J Sound Vib*. 2002;258: 405–417.
8. Mouri K, Fujii K, Shimokura R, Ando Y. A study on the dynamic properties of auditory interaural cross-correlation function relating to the moving sound image in the horizontal plane for the band noise. (Unpublished).
9. Schroeder MR. Natural sounding artificial reverberation. *J Audio Eng Soc*. 1962;10:219–223.
10. Ando Y, Shidara S, Maekawa Z, Kido K. Some basic studies on the acoustic design of room by computer. *JASJA*. 1973; 29:151–159 (in Japanese with English abstract).
11. Morimoto M, Ando Y. On the simulation of sound localization. *JASJA*. 1980;1:167–174.
12. Kirkeby O, Nelson PA, Hamada H. The "stereo dipole"—A virtual source imaging system using two closely spaced loudspeakers. *J Audio Eng Soc*. 1998;46:387–395.
13. Martignon P, Azzali A, Cabrera D, Capra A, Farina A. *Reproduction of Auditorium Spatial Impression with Binaural and Stereophonic Sound System*. Barcelona: Audio Engineering Society; 118th Convention, 2005.

CHAPTER 5

Magnetoencephalographic Evoked Responses to Factors Extracted from the Autocorrelation Function (ACF)/the Inter-Aural Cross-Correlation Function (IACF) Factors

MAGNETOENCEPHALOGRAPHIC RESPONSES TO EACH ACF FACTOR

Factor of the Sound Pressure Level

Auditory evoked magnetic field magnetoencephalographic (MEG) responses were recorded in relation to sound pressure level (SPL) and frequency (250 to 8000 Hz).[1] The auditory stimuli were tone bursts with 500 ms duration including rise and fall ramps of 10 ms. Evoked MEGs were recorded using a 122-channel whole-head magnetoencephalography system with 61 pairs of two orthogonally oriented planar gradiometers (Neuromag-122; Neuromag Lt., Helsinki, Finland) in a magnetically shielded room.

Typical and significant results of N1m-amplitude as a function of the SPL are shown in Fig. 5.1A for 500 Hz and Fig. 5.1B for 1000 Hz, respectively. Because with high frequencies greater than 2000 Hz the degree of amplitude decreased, the N1m (amplitude) might be expressed by

$$N1m_{amplitude} \sim f_R\left(SPL, W_{\phi(0)}\right) \qquad (5.1)$$

where $W_{\phi(0)}$ depends on the frequency component and is extracted from the ACF near the origin of delay time and is known as a cue of timbre.[2]

It is worth noting that $N1m_{amplitude}$ was higher in the right hemisphere than in the left, particularly at

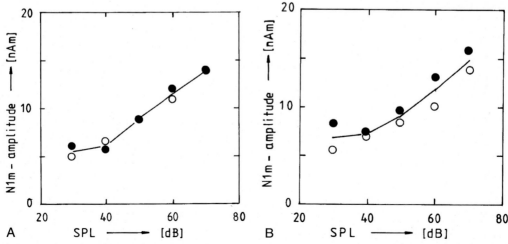

FIG. 5.1 N1m — amplitude (the standard error of means (SEM) ~ ±2.0) as a function of sound pressure level (SPL). (A) 500 Hz. (B) 1000 Hz. ○, Left hemisphere; ●, right hemisphere. (Soeta Y, Nakagawa S. Sound level-dependent growth of N1m amplitude with low and high-frequency tones. *NeuroReport*. 2009;20:548–552.)

1000 Hz.[3] This result agrees with results of the amplitude $A(P_1 - N_1)$ observed by the slow vertex response (SVR).[4]

Factors τ_1 and ϕ_1

Evoked MEGs were recorded in relation to pitch (τ_1) and pitch strength (ϕ_1). Iterated rippled noise was

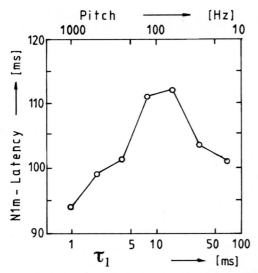

FIG. 5.2 N1m — latency (the standard error of means (SEM) ~ ±2.5) as a function of pitch period, τ_1. This figure was rearranged with reference to two articles.[5,6] The linear range between 1 and 16 ms (pitch frequency: 1000 Hz—62.5 Hz) is considered effective.

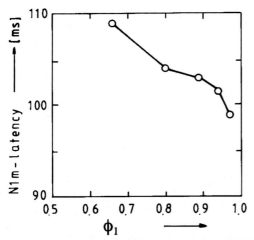

FIG. 5.3 N1m — latency (the standard error of means SEM ~ ±2.5) as a function of pitch strength, ϕ_1, extracted from the ACF, which was varied by the number of iterations with 2-ms delay. This figure was rearranged with reference to two articles.[5,6]

produced by a delay-and-add algorithm applied to band pass—filtered noise. The number of iterations of the delay-and-add process was set to 2, 4, 8, 16, and 32, which is related to the value of ϕ_1. In addition, the delay time was set to $\tau_1 = 1, 2, 4, 8, 16, 32,$ and 64 ms that corresponds to pitches of 1000, 500, 250, 125, 62.5, 31.3, and 15.6 Hz, respectively.

Results of the N1m$_{latency}$ increased with increasing delay time up to $\tau_1 = 8$—16 ms, as demonstrated in Fig. 5.2, and decreased with an increasing value of ϕ_1, as shown in Fig. 5.3. This figure was rearranged with reference to two articles.[5,6] Due to the previous two results of N1m$_{latency}$ corresponding to factors τ_1 and ϕ_1, we may approximately express

$$N1m_{latency} = f_L(\tau_1, \phi_1). \qquad (5.2)$$

Factor τ_e

To control the factor τ_e, the bandwidth of noise was adjusted by use of a sharp cut-off filter.[7] When the signal is sinusoidal with no bandwidth, the value of τ_e is infinite, and for white noise τ_e is zero.

Results of evoked MEG recordings are shown in Fig. 5.4, where N1m$_{ECD\ moment}$ increased with increasing τ_e value. N1m$_{ECD\ moment}$ varied more dynamically in the left hemisphere rather than the right.

This relation can be simply expressed, such that

$$N1m_{ECD\ moment} = f_L(\tau_e) \qquad (5.3)$$

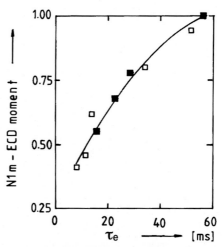

FIG. 5.4 N1m — Equivalent current dipole (ECD) moment (SEM < ±0.25) observed in the left hemisphere as a function of the effective duration of ACF. This figure was rearranged with reference to two source materials.[2,7] □, 500 Hz of the center frequency of sharply filtered white noise; ■, 1000 Hz of the center frequency of sharply filtered white noise.

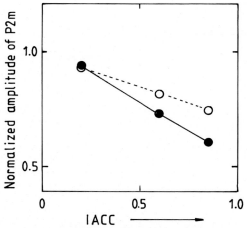

FIG. 5.5 Normalized amplitude of P2m (SEM < ±0.2) in relation to IACC. ○, Left hemisphere; ●, right hemisphere; *IACC*, interaural cross-correlation. (Soeta Y, Hotehama T, Nakagawa S, Tonoike M, Ando Y. Auditory evoked magnetic fields in relation to interaural cross-correlation of band-pass noise. *Hear Res.* 2004;196:109–114.)

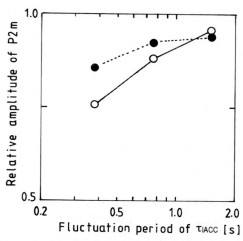

FIG. 5.6 Relative amplitude of P2m in relation to fluctuation period of τ_{IACC}. ○, Left hemisphere; ●, Right hemisphere; *IACC*, interaural cross-correlation. (Soeta Y, Nakagawa S, Tonoike. Auditory evoked fields to variations of interaural time delay. *Neurosci Lett.* 2005;383:311–316.)

TABLE 5.1
Evoked magnetoencephalographic (MEG) responses corresponding to each factor extracted from ACF/IACF.

(A) MEG CORRESPONDING TO EACH ACF FACTOR				
Signal used in the citations	ACF factors	MEG portions discovered in relation to the factor	Hemispheric dominance	Citations
Tone burst (500ms) of 250, 500,1000,…, 8000 Hz	**SPL** (loudness)	Amplitude of **N1m**	Right (1000Hz only)	Soeta and Nakagawa (2009), (2012).
Number of iterations with delay1,2,4,…64ms (pitch 1000, 500,…, 15.6 Hz)	τ_1(pitch)	Latency of **N1m**	—	Soeta and Nakagawa (2008a), (2008b).
Number of iterations with delay1,2,4,…64ms (pitch 1000, 500,…, 15.6 Hz)	ϕ_1(pitch strength)	Latency of **N1m**	—	Soeta and Nakagawa (2008a), (2008b).
Pure tone and bandpass noise with the center frequency of 500, 1000 and 2000 Hz	τ_e	ECD movement **N1m**	Left	Soeta, Nakagawa and Tonoike (2005); Fig. 15(c) in Ando, 2009.
(B) MEG CORRESPONDING TO EACH IACF FACTOR				
Signal used in the citations	IACF factors	MEG portions discovered in relation to the factor	Hemispheric dominance	Citations
Bandpass noise of 200–3000 Hz	**IACC**	Amplitude of **P2m**	Right[1]	Soeta, Hoehama, Nakagawa, Tonoike and Ando (2004).
Bandpass noise with the center frequency of 500 Hz	τ_{IACC} fluctuated	Amplitude of **P2m**	—[2]	Soeta, Nakagawa and Tonoike (2005).

[1] Dynamic range of both portions corresponding to IACC was much wider in the right hemisphere.
[2] The fluctuated signal was continuous and did not clearly determine right hemisphere specialization.

MEG RESPONSES TO IACF FACTORS

Factor Interaural Cross-Correlation

Evoked MEG responses of the human brain were analyzed in relation to the magnitude of interaural cross-correlation (IACC).[8]

The IACC of sound stimuli was controlled by mixing diotic band pass and dichotic independent band pass noise in certain ratio with identical SPL. The auditory stimuli were binaurally delivered through plastic tubes and earpieces inserted into ear canals with normal hearing (nine subjects).

Auditory evoked fields (AEFs) were recorded for combinations of a reference stimulus (IACC = 1.0), and test stimuli (IACC = 0.2, 0.6, or 0.85) were presented alternatively at a constant interstimulus interval of 0.5 s. Results are shown in Fig. 5.5. The normalized amplitude of P2m decreased with increasing IACC, so that

$$\text{Normalized P2m}_{\text{amplitude}} \sim f_R(\text{IACC}) \qquad (5.4)$$

The dynamic range of the values are more prominent in the right hemisphere. Note that the N1m$_{\text{latency}}$ and N1m$_{\text{amplitude}}$ were well related to temporal ACF factors; however, they were not affected by the spatial factor, IACC (Table 5.1).

Factor τ_{IACC}

Auditory motion in the horizontal direction may be simulated by presenting binaural sounds with time-varying interaural time delays.[9] AEFs were recorded for auditory motion from central to right and then to central. N1m$_{\text{amplitude}}$ and N1m$_{\text{latency}}$ were not affected by the fluctuation of interaural time delay; however, the relative amplitude of P2m significantly increased as a function of the interaural time delay as shown in Fig. 5.6. Thus it tentatively yields

$$\text{Relative P2m}_{\text{amplitude}} \sim f_R(\tau_{\text{IACC}}) \qquad (5.5)$$

In conclusion, results of the neural responses in relation to both ACF factors and IACF factors are listed in Table 5.1A and B, respectively.

REFERENCES

1. Soeta Y, Nakagawa S. Sound level-dependent growth of N1m amplitude with low and high-frequency tones. *NeuroReport.* 2009;20:548−552.
2. Ando Y. *Auditory and Visual Sensations.* New York: Springer-Verlag; 2009.
3. Soeta Y, Nakagawa S. Auditory evoked responses in human auditory cortex to the variation of sound intensity in an ongoing tone. *Hear Res.* 2012;287:67−75.
4. Ando Y. Evoked potentials relating to the subjective preference of sound fields. *Acustica.* 1992;76:292−296.
5. Soeta Y, Nakagawa S. The effects of pitch and pitch strength on an auditory-evoked N1m. *NeuroReport.* 2008;19: 783−787.
6. Soeta Y, Nakagawa S. Effect of repetitive components of a noise on loudness. *J Temporal Des Archit Environ.* 2008;8:1−7.
7. Soeta Y, Nakagawa S. Tonoike. Auditory evoked magnetic fields in relation to bandwidth variations of bandpass noise. *Hear Res.* 2005;202:47−54.
8. Soeta Y, Nakagawa S, Tonoike M. Magnetoencephalographic activities related to the magnitude of the interaural cross-correlation function (IACC) of sound fields. *J Temporal Des Archit Environ.* 2005;5:5−11.
9. Soeta Y, Nakagawa S. Tonoike. Auditory evoked fields to variations of interaural time delay. *Neurosci Lett.* 2005; 383:311−316.

SUGGESTED READING

Ando Y, Yamamoto K, Nagamatsu H, Kang SH. Auditory brainstem response (ABR) in relation to the horizontal angle of sound incidence. *Acoust Lett.* 1991;15:57−64.
Ando Y. Autocorrelation-based features for speech representation. *Acta Acust united Ac.* 2015;101:145−154.
Soeta Y, Hotehama T, Nakagawa S, Tonoike M, Ando Y. Auditory evoked magnetic fields in relation to interaural cross-correlation of band-pass noise. *Hear Res.* 2004;196:109−114.
Soeta Y, Nakagawa S. Tonoike. Auditory evoked magnetic fields in relation to bandwidth variations of bandpass noise. *Hear Res.* 2009;202:47−54.

CHAPTER 6

Neural Evidences Related to Subjective Preference

SLOW VERTEX RESPONSES RELATED TO SUBJECTIVE PREFERENCE OF SOUND FIELDS

Neuronal Response Correlates to Changes in Δt_1

First, subjective preferences were obtained by paired-comparison tests between different acoustic stimuli with different delay times for reflections. The preferred delay of the single reflection observed in these tests was typically in the 0 to 125 ms range according to the minimum effective duration $(\tau_e)_{min}$ of the running autocorrelation function of source signals. To compare slow vertex response (SVR) with these subjective preferences, a reference stimulus was first presented, and thereafter an adjustable test stimulus was presented. Typically, SVRs are at greatest value when stimulus patterns change abruptly (i.e., there is a contrast between the paired stimuli in which spatial and/or temporal factors change). The method of paired stimuli is therefore the most effective procedure because of this relativity of the brain response.

Electrical responses were obtained from the left and right temporal areas (T3 and T4) according to the International 10 to 20 System.[1] The reference electrodes were located on the right and left earlobes and were joined together. Each subject had been asked to abstain from smoking and from drinking alcohol for 12 hours prior to the experiment. Such pairs of stimuli were presented alternately 50 times through two loudspeakers. The two loudspeakers were located directly in front of the subject, $(68 \pm 1$ cm$)$, so that the magnitude of inter-aural cross-correlation (IACC) could be kept at a constant value near unity for all stimuli. The SVR for each trial was integrated and averaged as with other auditory evoked potentials (AEPs).

Fig. 6.1 shows examples of such averaged SVRs, as a function of the delay time of the single sound reflection, Δt_1, a sound delayed and added to itself. The source signal was a 0.9 s fragment of continuous speech, namely, the Japanese word, "ZOKI-BAYASHI," which means "grove" or "thicket." The reference "direct" stimulus

sound field was the source signal without any added reflections, whereas the second "reflection" stimulus was the source signal delayed and added to itself. The amplitude of the reflection was the same as that of the direct sound $A_0 = A_1 = 1$, with delays ranging between 0 and 125 ms. Total sound pressure levels were kept constant, precisely between the direct and reflected sound stimuli by measuring the ACF, $\Phi_p(0)$. It is obvious that the amplitude from the left hemisphere was always greater than from the right, $P < .01$.

From the center column in Fig. 6.4, for example, we found the longest neuronal response latencies (for the P_2 wave this is the rightmost extent of the dotted line) for the stimulus that has the most preferred reflection time delay (25 ms). This indicates the most comfortable (i.e., the maximally preferred) delay for listening to continuous speech and allowing the mind to relax. The delay time of 25 ms corresponds to the effective duration of the ACF, τ_e, of a continuous speech signal.[2]

Hemispheric Response Difference to Change in Listening Level and Interaural Cross-correlation

Because our environment always consists of temporal factors as well as spatial factors, we assume that different interhemispheric responses may be found according to the two classified factors. Analogous hemispheric differences in the initial SVR amplitude were observed as a function of sensation level (SL) and the magnitude of IACC. For all levels greater than 30 dB SL, the initial SVR amplitude from the right hemisphere was greater than from the left, $P < .01$ (Fig. 6.2).[3] We used $1/3$-octave band noise with center frequency 500 Hz to examine the response correlates of sound direction. For all IACC values from 0.1 to 1.0 in the paired stimuli, the amplitude from the right hemisphere was always greater than from the left, $P < .01$ (Fig. 6.3).[4] By assembling the data (Table 6.1), we can see that hemispheric dominance of the SVR responses can change as a function of the

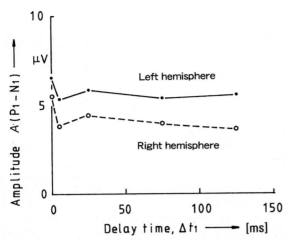

FIG. 6.1 Averaged amplitudes A(P₁ − N₁) of the test sound field over the left and right hemispheres, in change of delay time of reflection, Δt₁ (eight subjects). (⸻), Left hemisphere; (− − − −), right hemisphere.

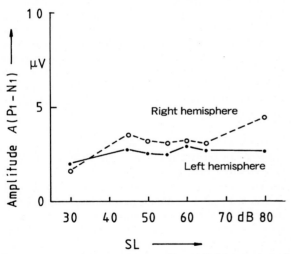

FIG. 6.2 Averaged amplitudes, A(P₁ − N₁) of the test sound field over the left and right hemispheres, in change of sensation level (SL) (five subjects). (⸻), Left hemisphere; (− − − −), right hemisphere.

variation of the acoustic parameter. Remarkably, when the spatial factor IACC was varied in the paired stimuli, the right hemisphere was highly activated. In addition, the right hemisphere was dominant under the condition of a varying SL, even if the continuous speech signal α was used as the source signal. On the other hand, the left hemisphere was dominant under the condition of a varying Δt₁, which is a temporal factor of the sound field. From classic studies, the left

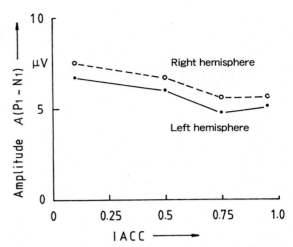

FIG. 6.3 Averaged amplitudes, A(P₁ − N₁) of the test sound field over the left and right hemispheres in change of IACC (eight subjects). (⸻), Left hemisphere; (− − − −), right hemisphere; *IACC*, interaural cross-correlation.

hemisphere appears to be more highly involved with processing speech and temporal sequences, whereas the right is concerned with nonverbal and spatial pattern identifications.[5,6] In light of our results, some aspects of hemispheric dominance may be relative to phenomena that are dependent on what is changed in the pair of stimuli (i.e., temporal vs. spatial factors). These results discard the possibility of absolute conditions and motivate further studies to individualize sound fields.

Differences in Response Latency Corresponding to Subjective Preference

We found neural response correlates of subjective preference in the latency of SVR waves. The top plots of Fig. 6.4 summarize the relationship between subjective preference scale values and three acoustic parameters (SL, Δt₁, and IACC). Applying the paired method of stimuli, both SVR and the subjective preference for sounds fields were investigated as functions of SL and Δt₁. The source signal was a 0.9-s speech segment. The lower part of this figure indicates the appearance of latency components. As shown in the left and center columns in this figure, the neural information related to subjective preference appeared typically in an N₂-latency of 250 to 300 ms, when SL and Δt₁ were changed.

Further details of the latencies for both the test sound field and the reference sound field, when Δt₁ was changed, are shown in Fig. 6.5. The parallel latencies at P₂, N₂, and P₃ were clearly observed as

TABLE 6.1

Scale Value of Subjective Preference as a Function of the four Orthogonal Factors of the Sound Field. According to the Subjective Preference Theory, the Total Scale Value of the Sound Field is Expressed by (Ando, 1998, 2007) $S = S_1 + S_2 + S_3 + S_4$, where $S_i \sim -\alpha|x_i|^{3/2}$, $i = 1, 2, 3, 4$ (Ando, 1985, 1998, 2007).

Factors changed	i	x_i	α $x_i > 0$	α $x_i < 0$
TEMPORAL				
Δt_1	2	$\log(\Delta t_1/[\Delta t_1]_p)$	1.42	0.04
T_{sub}	3	$\log(T_{sub}/[T_{sub}]_p)$	0.45 +0.75A	2.36 − 0.42
SPATIAL				
LL	1	$20\log P - 20\log[p]_p$ (dB)	0.07	0.04
IACC[1]	4	IACC	1.45	—

[1]Interaural time difference τ_{IACC} should be zero, the condition of frontal direction of the sound source.

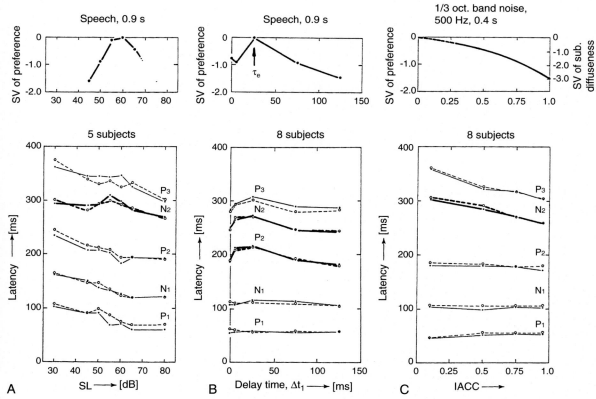

FIG. 6.4 Relationships between averaged latencies of slow vertex response and scale values of subjective preference for three factors of the sound field. (____), Left hemisphere; (− − − −), right hemisphere. (a) As a function of sensation level (SL). (b) As a function of delay time of reflection, Δt_1. (c) As a function of IACC. SVR, Slow vertex response; SL, sensation level; IACC, interaural cross-correlation.

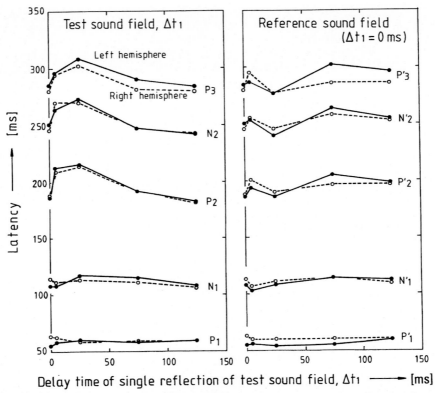

FIG. 6.5 Averaged latencies for the test sound field and the reference sound field for paired stimuli, as a function of delay time of reflection, Δt_1. (——): Left hemisphere; (————): Right hemisphere. Maximum latencies of P_2, N_2, and P_3 are found at $\Delta t_1 = 25$ ms for the test sound field, whereas relatively short latencies of P'_2, N'_2 and P'_3 are observed for the reference sound field. This is typical brain activity showing "relativity."

functions of the delay time, Δt_1. Latencies for the reference sound field ($\Delta t_1 = 0$) in the paired stimuli should be constant due to $\Delta t_1 = 0$ ms. However, the latencies were found to be relatively shorter, whereas the latencies for the test sound field with $\Delta t_1 = 25$ ms were the longest. This may indicate a kind of *"relative"* behavior of the brain, underestimating the reference sound field when the test sound field in the pair is the most preferred condition.

Relatively long-latency responses are always observed in the subjectively preferred range of each factor. Thus the difference of N_2-latencies over both hemispheres in response to a pair of sound fields contains almost the same information as was obtained from paired-comparison tests for subjective preference. Again, subjective preference may be summarized as judgments made in the direction of maintaining life; therefore it may appear in neuronal responses as a primitive response.

The right column of Fig. 6.4 shows the effects of varying IACC using $^1/_3$-octave band noise (500 Hz).[4] In the upper part, the scale value of subjective diffuseness is indicated as a function of IACC. The scale value of subjective preference behaved similarly plotted against the IACC, when speech or music signals were presented. The information related to subjective diffuseness or subjective preference, therefore, appears in the N_2-latency, ranging from 260 to 310 ms, in which a tendency for an increasing latency while decreasing the IACC was observed for eight subjects (except for the left hemisphere of one subject). As already indicated in Fig. 6.4 (c), the relationship between the IACC and the N_2-latency was found to be linear and the correlation coefficient between them was -0.99 ($P < .01$).

Furthermore, let us look at the behavior of early latencies of P_1 and N_1. These were almost constant when delay time and IACC were changed. However, information related to SL or loudness may be typically found at the N_1-latency. This tendency agrees well with the results of Botte, Bujas, and Chocholle (1975).[7]

Consequently, from 40 to 170 ms of the SVR, hemispheric dominance may be found for the amplitude component as shown in Figs. 6.1–6.3, which may indicate respective functional specialization of the hemispheres. Early latency differences corresponding to SL may be found in the range from 120 to 170 ms. Finally, we found that the N_2-latency components in the delay range between 200 and 310 ms may correspond well with subjective preference relative to listening level, time delay of the reflection, and, indirectly, the IACC. Because the longest latency was always observed for the most preferred condition, one might speculate that the brain is at its most relaxed state at the preferred condition and that this causes the observed latency behavior to occur. This extended period may relate to the α-wave having the longest period in electroencephalography (EEG) and magneto-encephalography (MEG) during the human waking stage, as discussed later.

We discussed in Chapter 2 of section "Auditory Brainstem Responses in Auditory Pathways" that the activity of the ABR in the short delay range (<10 ms) after the sound signal has arrived at the eardrums may reflect possible mechanisms in the auditory pathway for analyzing interaural correlation patterns (Fig. 2.18).

Electroencephalographic and Magnetoencephalographic Response Correlates of Subjective Preference

To attain further knowledge on brain activities, we investigated continuous EEG response correlates of subjective preference, covering changes in reverberation time (usually, 0.5 s < T_{sub} < 5.0 s) for continuous test signals that could not be studied with evoked responses where only short signals less than 0.9 s can be applied. First of all, we aimed at finding a distinctive feature in EEG by changing the delay time of a single reflection (Δt_1) to reconfirm the SVR results. To obtain individual differences more clearly, we conducted further investigation also on MEG responses following changes to Δt_1. Finally, we studied EEG responses to changes in T_{sub} and the effects of the typical spatial factor (IACC) through EEG recordings.

Electroencephalography in Response to Change in Δt_1

In this experiment, music motif B (Arnold's Sinfonietta, Opus 48, a 5-s segment of the third movement) was selected as the sound source.[8,9] The delay time of the single reflection, Δt_1, was alternately adjusted to 35 ms (a preferred condition) and 245 ms (a condition of echo disturbance). The EEG of 10 pairs from T_3 and T_4 was recorded for approximately 140 s, and experiments were repeated over a total of 3 days. Eleven 22- to 26-year-old subjects participated in the experiment. Each subject was asked to close his or her eyes when listening to the music during the EEG recording session. Two loudspeakers were arranged in front of the subject, keeping the IACC at a constant value near unity. The sound pressure level was fixed at a 70-dBA peak, in which the amplitude of the single reflection was the same as that of the direct sound, $A_0 = A_1 = 1$. The leading edge of each sound signal was recorded at the same time for EEG analysis. The EEG recording was sampled at greater than 100 Hz after passing through a filter width of 5 to 40 Hz with a slope of 140 dB/octave.

To find brain activity corresponding to subjective preference, we attempted to analyze the effective duration of the ACF, τ_e in the α-wave range (8 to 13 Hz) of the EEG. First, considering that a subjective preference judgment needs at least 2 s to develop a psychological present, the running integration window (2T) was examined for periods between 1.0 and 4.0 s.

1. A satisfactory duration 2T = 2 to 3 s in the ACF analysis was found only from the left hemisphere but not from the right.[10]
2. To analyze the data in more detail for each category, Δt_1 and left and right cerebral hemispheres (LR), we show the averaged value of τ_e in the α-wave from 11 subjects in Fig. 6.6. A clear tendency is apparent. Values of τ_e at Δt_1 = 35 ms are significantly longer than those at Δt_1 = 245 ms ($P < .01$) only on the left hemisphere, not on the right.
3. Ratios of τ_e values in the α-wave range at Δt_1 = 35 ms and 245 ms, for each subject, are shown in Fig. 6.7. Remarkably, all individual data indicate that the ratios in the left hemisphere at the preferred condition of 35 ms are much longer than in the right hemisphere.
4. Thus results reconfirm that, when Δt_1 is changed, the left hemisphere is highly activated, and the value of τ_e for the α-wave of this hemisphere corresponds well to subjective preference.

It is worth noting that the α-wave has the longest period in the EEG in the waking state and may indicate feelings of "pleasantness" (i.e., a preferred condition which is widely accepted). Thus a long value of τ_e in the α−wave may relate to the long N_2-latency in SVR at the preferred condition as shown in Figs. 6.4 and 6.5.

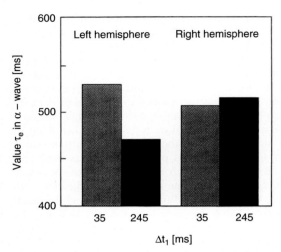

FIG. 6.6 Averaged ACF τ_e values of the electroencephalography-α-wave in change of Δt_1: 35 ms and 245 ms (11 Subjects). Left, Left hemisphere; Right, right hemisphere. Significant difference of ACF τ_e values may be found over the left hemisphere, but not on the right. *EEG*, Electroencephalography.

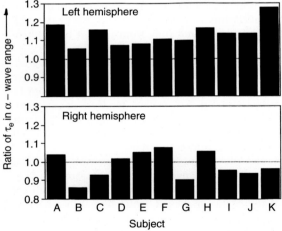

FIG. 6.7 Ratios of ACF τ_e value of the electroencephalography (EEG)-α-wave of 35 ms and 245 ms in change of Δt_1 for each individual subject, A–K, [τ_e value at 35 ms] / [τ_e value at 245 ms]. Above, left hemisphere; Below, right hemisphere. Ratios of ACF τ_e values of the EEG-α-wave are always greater on the left hemisphere than on the right.

Magnetoencephalography in Response to Change in Δt_1

In MEG studies, weak magnetic fields produced by electric currents flowing in neurons are measured with multiple channel SQUID (superconducting quantum interference device) gradiometers, which enable the study of many interesting properties of the working human brain. MEG accurately detects superficial tangential currents, whereas EEG is sensitive to both radial and tangential current sources and also reflects activity in the deepest parts of the brain. Only currents that have a component tangential to the surface of a spherically symmetric conductor produce a sufficiently strong magnetic field outside of the brain; radial sources are thus externally silent. Therefore MEG measures mainly activity from the fissures of the cortex, which often simplifies interpretation of the data.

Fortunately, all the primary sensory areas of the brain—auditory, somatosensory, and visual—are located within fissures. The advantages of MEG over EEG result mainly from the fact that, although the skull and other extracerebral tissues substantially alter current flow, they are practically transparent to magnetic fields. Thus magnetic patterns outside the head are less distorted than the electrical potentials on the scalp. Furthermore, magnetic recording is reference free, whereas electric brain maps depend on the location of the reference electrode.

Measurements of MEG responses were performed in a magnetically shielded room using a 122-channel whole-head neuromagnetometer, as shown in Photo 6.1. (Neuromag-122, Neuromag Ltd., Finland).[11]

The source signal was the word, "piano," with a 0.35-s duration. The minimum value of the moving τ_e (i.e., $[\tau_e]_{min}$) was approximately 20 ms. It is worth noting that this value is close to the most preferred delay time of the first reflection of sound fields with continuous speech.[12] Knowledge of preferred reflections may improve comfort of hearing offered by a well-designed individual hearing aid, for example. In the present experiment, the delay time of the single reflection (Δt_1) was set at five levels (0, 5, 20, 60, and 100 ms). The direct sound and a single reflection were mixed and the amplitude of the reflection was the same as for the direct sound ($A_0 = A_1 = 1$). The auditory stimuli were binaurally delivered through plastic tubes and earpieces into the ear canals without any metallic material. The sound-pressure level, which was measured at the end of the tubes, was fixed at 70 dBA.

Eight 23- to 25-year-old subjects participated in the experiment. All had normal hearing. In accordance with the paired comparison test (PCT), each subject compared 10 possible pairs per session, and a total of 10 sessions were conducted for each subject. Measurements of magnetic responses were performed in a magnetically shielded room. Similar to previous EEG measurements, paired-auditory stimuli were presented

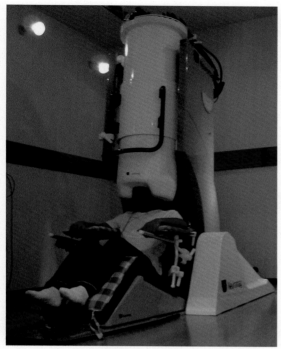

PHOTO 6.1 Magnetometer used in recording the magnetoencephalogram (MEG).

in the same way as in the subjective preference test. During measurements, the subjects sat in a chair with their eyes closed. To compare the results of the MEG measurements with the scale values of subjective preference, combinations of a reference stimulus ($\Delta t_1 = 0$ ms) and test stimuli ($\Delta t_1 = 0, 5, 20, 60,$ and 100 ms) were presented alternately 50 times, and the MEGs were analyzed. The magnetic data were recorded continuously with a filter of 0.1 to 30.0 Hz and digitized at a sampling rate of 100 Hz. Eight channels that had a larger amplitude of N1m response in each hemisphere were selected for the ACF analysis. We analyzed the MEG α-wave for each of the paired stimuli for each subject. The value of τ_e is defined by the delay time at -10 dB from the origin (0 dB) of the normalized ACF (namely, ten percentile delay). It is practically be extrapolated at the delay time by the straight line at -5 dB from the top (0 dB) of the normalized ACF expressed in the dB scale. Obviously, for the preferred condition at $\Delta t_1 = 5$ ms of the sound field, the value of $\tau_e \approx 0.5$ s, and for the condition of echo disturbance ($\Delta t_1 = 100$ ms), the value of $\tau_e \approx 0.3$ s, which is much smaller.

The results from the eight subjects confirm a linear relationship between the averaged τ_e values of α-wave and the averaged scale values of subjective preference.

Because the left hemisphere dominates Δt_1, reconfirming the aforementioned studies of SVR and EEG, we analyzed the results from the left hemisphere on individual level.

1. We found an almost direct relationship between individual scale values of subjective preference and τ_e values over the left hemisphere in each of the eight subjects. Results for each subject are shown in Fig. 6.8.
2. Remarkably, the correlation coefficient achieved between the scale values of subjective preference and τ_e values over the left hemisphere, r, was more than 0.94 for all subjects.
3. However, there is a very weak relationship between the scale values of subjective preference and the amplitude of α-wave, $\Phi(0)$, in both hemispheres ($r < 0.37$).
4. The value of τ_e is the degree of similar repetitive features apparent in α-waves, so that the brain repeats a similar rhythm under the preferred conditions. This tendency for a larger τ_e under the preferred condition is much more significant than the results of EEG α-waves mentioned previously.

Electroencephalography in Response to Change in T_{sub}

Now, let us examine values of τ_e in the α-wave with changes to the subsequent reverberation time (T_{sub}) relative to the scale values of subjective preference.[13] Ten student subjects participated in the experiment.[13] The sound source used was music motif B, described with $(\tau_e)_{min} \sim 40$ ms, so that the most preferred reverberation time calculated is $(T_{sub})_p \sim 23\ (\tau_e)_{min} = 0.92$ s. Ten 25- to 33-year-old subjects participated in the experiment. The EEG from the left and right hemisphere was recorded. Values of τ_e of the α-wave, for the duration, $2T = 2.5$ s, were also analyzed here.

First, consider the averaged values of τ_e of the α-wave, shown in Fig. 6.9.

1. Clearly, the values of τ_e are much longer at close to the preferred condition 0.92 s, $T_{sub} = 1.2$ s than at $T_{sub} = 0.2$ s in the left hemisphere, whereas the values of τ_e are longer at $T_{sub} = 1.2$ s than at $T_{sub} = 6.4$ s.
2. It is significant in the left hemisphere. However, this is not true for the right hemisphere.

The results of analysis of variance are that, although there are large individual differences, a significant difference is achieved for T_{sub} in the pair of 0.2 s and 1.2 s ($P < .05$), and interference effects are observed for factors Subject and LR ($P < .01$), and LR and T_{sub} ($P < .01$). No such significant differences are achieved for the pair at 1.2 s and 6.4 s, but there are interference effects between Subject and LR, and Subject and T_{sub}. However, we have discussed that differences in scale

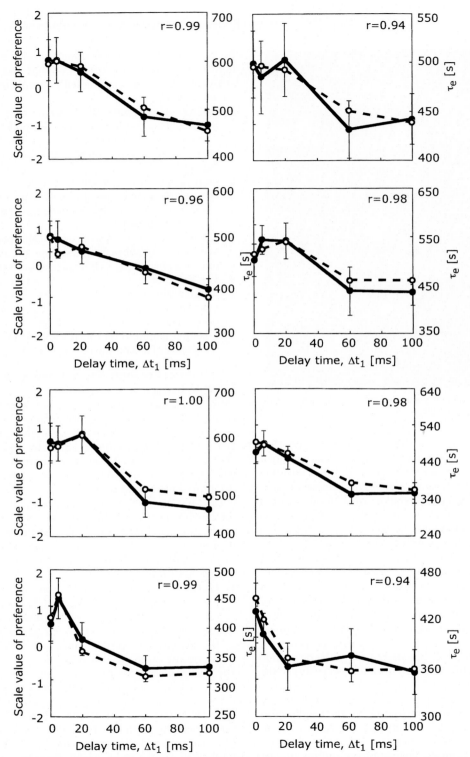

FIG. 6.8 Good correspondence between the individual scale value of subjective preference and averaged ACF τ_e value of the magnetoencephalography (MEG)-α-wave over the individual left hemisphere (eight subjects). The averaged τ_e value and the scale value are in highest correlation over the eight channels. $\bigcirc$, Scale values of subjective preference; $\bullet$, averaged τ_e values of MEG α-wave, error bars are standard errors.

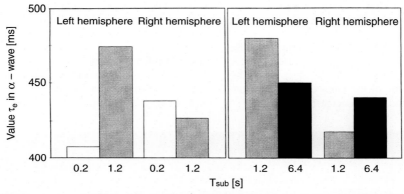

FIG. 6.9 Averaged value of ACF τ_e of the electroencephalography-α-wave in change of T_{sub}: 0.2 s and 1.2 s; 1.2 s and 6.4 s, for 10 subjects. Left bars in each figure represent the left hemisphere, right bars represent the right hemisphere.

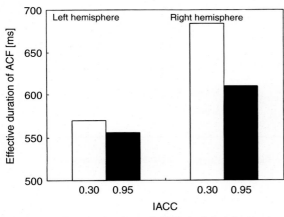

FIG. 6.10 Averaged values of ACF τ_e of the electroencephalography-α-wave in change of interaural cross-correlation (IACC) for the pair of IACC = 0.30 and 0.95. Left bars represent the left hemisphere, right bars represent the right hemisphere.

values of preference correspond well to ratios in α-waves in the left hemisphere.[13-15]

Electroencephalography in Response to Change in Interaural Cross-correlation

We also investigated EEG response to changes in IACC. Eight student subjects participated in the experiment.[16] More clearly here, with changes to IACC using music motif B, right-hemisphere dominance is shown using analysis of the value of τ_e for the α-wave. A significant difference is achieved in the right hemisphere for the pair of sound fields of IACC = 0.95 and 0.30 ($P < .01$) in the results shown in Fig. 6.10.

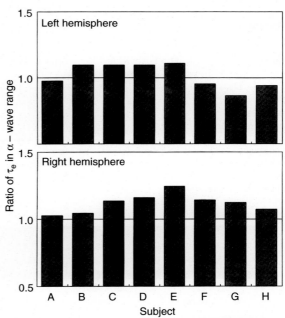

FIG. 6.11 Ratio of ACF τ_e values of the electroencephalography-α-wave from the left hemisphere (T3) and the right hemisphere (T4) for each of eight subjects. A–H, [τ_e value at IACC = 0.30]/[τ_e value at interaural cross-correlation = 0.95]. Ratio of ACF τ_e values are greater on the right hemisphere than on the left except for subject B.

1. The ratio of the values of τ_e for the α-wave for seven of eight subjects with change in IACC, [τ_e (IACC = 0.3) / τ_e(IACC = 0.95)], in the right hemisphere are greater than in the left hemisphere except for subject B (Fig. 6.11). Thus, as far as the IACC is concerned, the

more preferred condition with a smaller IACC is related to a longer value of τ_e for the α-wave in the right hemisphere in most tested subjects.

2. We also reconfirmed by MEG experiments with the same speech signal when changing the IACC (0.27, 0.61, and 0.90) that the values of τ_e and the maximum amplitude of the cross correlation function (CCF) increased when decreasing the IACC.[17]

As listed in Table 6.1, the scale value of four orthogonal factors of the sound field may be calculated. The table summarizes hemispheric dominance obtained by analysis of τ_e in α-waves, with changes in LL, Δt_1, T_{sub}, and IACC (Table 6.2). Scale values of subjective preference for the four orthogonal factors of the sound field with speech and music that clearly appeared in activities of AEP, EEG, and MEG are listed in Table 6.3.

TABLE 6.2
Hemispheric Specializations of Temporal and Spatial Factors of the Sound Field Determined by Analyses of the Auditory Evoked Potentials (AEP), EEG and MEG (Ando, 2009).

Factors changed	AEP (SVR) $A(P_1 - N_1)$	EEG, ratio of ACF τ_e values of α-wave	AEP (MEG) N1m	MEG, ACF τ_e value of α-wave
TEMPORAL				
Δt_1	**L > R** (speech)[1]	**L > R** (music)	**L > R** (speech)	
T_{sub}	—	**L > R** (music)	—	
SPATIAL				
LL	**R > L** (speech)	—	—	
IACC	**R > L** (vowel /a/)	**R > L** (music)[2]	**R > L** (band noise)[3]	
	R > L (band noise)			
τ_{IACC}			**R > L** (band noise)[3]	
Head related transfer functions			**R > L** (vowels)[4]	

[1]Sound source used in experiments is indicated in the bracket.
[2]Flow of EEG α-wave from the right hemisphere to the left hemisphere for music stimulus in change of the IACC was determined by the CCF $|\phi(\tau)|_{max}$ between α-waves recorded at different electrodes.
[3]Soeta and Nakagawa, 2006.
[4]Palomaki, Tiitinen, Makinen, May and Alku, 2002.

TABLE 6.3
Scale Values of Subjective Preference of Four Orthogonal Factors of the Sound Field with Speech or Music that Appeared in Activities of AEP, EEG and MEG (Ando, 2009).

Factors changed	AEP (SVR) Latency	EEG, ratio of ACF τ_e in α-wave	MEG, ACF τ_e value of α-wave
TEMPORAL			
Δt_1	N_2 and $P_2 \sim SV_P$ Figure 6.4	$(\tau_e$ in α-wave)$_L \sim SV$ Figure 6.6	$(\tau_e$ in α-wave)$_L = SV$ Figure 6.8
T_{sub}		$(\tau_e$ in α-wave)$_L \sim SV$ Figure 6.9	
SPATIAL			
LL	$N_2 \sim SV_P$ Figure 6.4		
IACC	$N2 \sim SV_D$[1] Figure 6.4	$(\tau_e$ in α-wave)$_R \sim SV$ Figure 6.10	

[1]Sound source used in subjective diffuseness experiments was the 1/3 oct. band noise, 500 Hz.

This finding may indicate that the values of τ_e in the α-range of brain waves are objective indices for maximizing scale value as a preferred condition when designing the human environment.[18]

SPECIALIZATION OF CEREBRAL HEMISPHERES FOR TEMPORAL AND SPATIAL FACTORS

As described previously, important evidences are summarized in Table 6.2,[19] which have been discovered by ABR, SVR, EEG, and MEG recordings over the left and right cerebral hemispheres.

1. The left and right amplitudes of the early SVR, $A(P_1 - N_1)$, indicate that left and right dominance occurs due to the temporal factor (Δt_1) and spatial factors (LL and IACC), respectively. It is worth noting that the SL or LL was first thought to be classified as a temporal-monaural factor from a physical viewpoint. However, results from SVR indicated that sound level is right hemisphere dominant. Thus SL or LL should be classified as a spatial factor from the viewpoint of the brain, which is measured by the geometric average value of the binaural sound energies arriving at both ears.

2. Both left and right latencies of N_2 correspond well to the values of IACC.

3. EEG results for the cerebral hemispheric specialization of the temporal factors, Δt_1 and T_{sub}, indicated left hemisphere dominance, whereas the IACC was a right hemispheric factor. There is thus a high degree of independence between the left and right hemispheres concerning judgment of subjective attributes, which we discussed in section "Slow Vertex Responses (SVR) Related to Subjective Preference of Sound Fields."[9]

4. When Δt_1 was changed, amplitudes of recorded MEG reconfirmed left hemisphere specialization.

5. The scale value of subjective preference corresponded well to the value of τ_e extracted from the ACF of the α-wave over the left and right hemispheres according to changes in temporal and spatial factors of sound fields, respectively.

6. The scale values of individual subjective preference related directly to the value of τ_e extracted from the auto correction function (ACF) of the MEG α-wave.

7. In addition to the previously mentioned activities on both the left and right hemispheres, spatial activities in the brain were analyzed by the cross-correlation function of EEG and MEG α-waves. Results showed that a large area of the brain was activated when preferred sound fields were presented.[15] This evidence implies that the brain repeats a similar temporal rhythm in the α-wave range throughout the area over the scalp under the preferred conditions of sound fields.

It is worth noting that similar findings have been evident from investigations in visual fields.[16,20,21] We reconfirm here that the left hemisphere is mainly associated with speech and time-sequential identifications, and the right is concerned with nonverbal and spatial identification.[5,6] However, when we changed IACC using speech and music signals, right hemisphere dominance was always observed. Therefore hemispheric dominance is of "relative" nature, depending on which factor is changed in the comparison pair, and no absolute behavior could be observed.

We have discovered that LL and IACC are dominantly associated with the right cerebral hemisphere, and the temporal factors, Δt_1 and T_{sub}, of the sound field in a room are associated with the left. Thus both temporal and spatial factors for any sound environment (e.g., concert halls, lecture rooms, churches, and conference rooms, or even hearing aids) should be taken into account when creating spaces to satisfy both hemispheres. It is remarkable that we can easily explain "cocktail party effects" in terms of specialization of the human brain because speech signals are processed in the left hemisphere and independently the directional information of a target speaker is mainly processed in the right hemisphere.

REFERENCES

1. Jasper HH. The ten-twenty electrode system of the international federation. *Electroencephalogr Clin Neurophysiol.* 1958;10:370–375.
2. Ando Y, Kang SH, Morita K. On the relationship between auditory-evoked potential and subjective preference for sound field. *JASJA(E).* 1987;8:197–204.
3. Nagamatsu H, Kasai H, Ando Y. Relation between auditory evoked potential and subjective estimation—effect of sensation level. *Rep Meeting Acoust Soc Jpn.* 1989:391–392 (in Japanese).
4. Ando Y, Kang SH, Nagamatsu H. On the auditory-evoked potentials in relation to the IACC of sound field. *JASJA(E).* 1987;8:183–190.
5. Kimura D. The asymmetry of the human brain. *Scientific American.* 1973;228:70–78.
6. Sperry RW. Lateral specialization in the surgically separated hemispheres. In: Schmitt FO, Worden FC, eds. *The Neurosciences: Third study program.* Cambridge: MIT Press; 1974, Chapter 1.
7. Botte MC, Bujas Z, Chocholle R. Comparison between the growth of averaged electroencephalic response and direct loudness estimations. *J Acoust Soc Am.* 1975;58:208–213.

8. Burd AN. Nachhallfreir Musik fuer akustische Modell-untersuchungen. *Rundfunktechn Mitteilungen.* 1969;13:200—201.

9. Ando Y. *Concert Hall Acoustics.* Heidelberg: Springer-Verlag; 1985.

10. Ando Y, Chen C. On the analysis of the autocorrelation function of α-waves on the left and right cerebral hemispheres in relation to the delay time of single sound reflection. *J Archit Plan Environ Eng Archit Inst JPN (AIJ).* 1996;488:67—73.

11. Soeta Y, Nakagawa S, Tonoike M, Ando Y. Magnetoence-phalographicresponses corresponding to individual subjective preference of sound fields. *J Sound Vib.* 2002;258:419—428.

12. Ando Y, Kageyama K. Subjective preference of sound with a single early reflection. *Acustica.* 1977;37:111—117.

13. Chen C, Ando Y. On the relationship between the autocor-relation function of the α-waves on the left and right cerebral hemispheres and subjective preference for the reverberation time of music sound field. *J Archit Plan Environ Engin, Archit Inst JPN (AIJ).* 1996;489:73—80.

14. Ando Y. *Architectural Acoustics, Blending Sound Sources, Sound Fields, and Listeners.* New York: AIP Press/Springer-Verlag; 1998.

15. Ando Y. *Auditory and Visual Sensations,* ed. Cariani P. New York: Springer-Verlag; 2009.

16. Sato S, Nishio K, Ando Y. Propagation of alpha waves corresponding to subjective preference from the right hemisphere to the left with change in the IACC of a sound field. *J Temporal Des Archit Environ.* 2003;3:60—69.

17. Soeta Y, Nakagawa S, Tonoike M. Magnetoencephalo-graphic activities related to the magnitude of the interaural cross-correlation function (IACC) of sound fields. *J Temporal Des Archit Environ.* 2005;5:5—11. http://www.jtdweb.org/journal/.

18. Ando Y. *Brain-Grounded Theory of Temporal and Spatial Design in Architecture and the Environment.* Tokyo: Springer; 2016.

19. Ando Y. *Brain Oriented Acoustics.* Tokyo (In Japanese): Itto-Sha; 2011.

20. Okamoto Y, Soeta Y, Ando Y. Analysis of EEG relating to subjective preference of visual motion stimuli. *J Temporal Des Archit Environ.* 2003;3:36—42. http://www.jtdweb.org/journal/.

21. Soeta Y, Nakagawa S, Tonoike M, Ando Y. Spatial analysis of magnetoencephalographic alpha waves in relation to subjective preference of a sound field. *J Temporal Des Archit Environ.* 2003;3:28—35. http://www.jtdweb.org/journal/.

SUGGESTED READING

Palomaki K, Tiitinen H, Makinen V, May P, Alku P. Cortical processing of speech sounds and their analogues in a spatial auditory environment. *Cogn Brain Res.* 2002;14:294—299.

Soeta Y, Nakagawa S. Auditory evoked magnetic fields in relation to interaural time delay and interaural crosscorrelation. *Hear Res.* 2006;220:106—115.

CHAPTER 7

Temporal and Spatial Primary Percepts of Sound and the Sound Field

TEMPORAL PRIMARY PERCEPTS IN RELATION TO TEMPORAL FACTORS OF A SOUND SIGNAL

Pitch of Complex Tones

We will show that factor τ_1 extracted from the autocorrelation function (ACF) of a sound signal directly describes the pitch period. It is remarkable that harmonic complexes having no energy at the fundamental frequency in their power spectra can still produce strong "low" pitches at the fundamental itself. It is thus the case for complex tones with a "missing fundamental" that strong pitches are heard which correspond to no individual frequency component, and this raises deep questions about whether patterns of pitch perception are consistent with frequency-domain representations.

A pitch-matching test comparing pitches of pure and complex tones was performed to reconfirm previous results.[1] The test signals were all complex tones consisting of harmonics 3 to 7 of a 200-Hz fundamental. All tone components had the same amplitudes, as shown in Fig. 7.1. As test signals, the two waveforms of complex tones, (1) in-phase and (2) random-phase, were applied as shown in Fig. 7.2. Starting phases of all components of the in-phase stimuli were set at zero. The phases of the components of random-phase stimuli were randomly set to avoid any periodic peaks in the real waveforms. As shown in Fig. 7.3, the normalized ACF of these stimuli were calculated at the integrated interval $2T = 0.8$ seconds. Although the waveforms differ greatly from each other, as shown in Fig. 7.2, their ACF are nevertheless experimentally and theoretically identical. The time delay at the first maximum peak of the ACF, τ_1, equals 5 ms (200 Hz), corresponding well to the fundamental frequency. The subjects were five musicians (two males and three females, 20 to 26 years of age). Test signals were produced from a loudspeaker in front of the subject in a semianechoic chamber. The sound pressure level (SPL) of each complex tone at the center position of the listener's head was fixed at 74 dB by analysis of the ACF $\Phi(0)$. The distance between a subject and the loudspeaker was 80 ± 1 cm.

The probabilities of matching frequencies counted for each $1/12$-octave band (chromatic scale) of the in-phase stimuli and random-phase stimuli are shown in Fig. 7.4. The dominant pitch of 200 Hz is included neither in the spectrum nor in the real waveform of random phases. However, it is obviously included in the period of the ACF. For both in-phase and random-phase conditions, approximately 60% of the responses clustered within a semitone of the fundamental. Results obtained for pitch under the two conditions are definitely similar. In fact, the pitch strength remains the same under both conditions. Thus pitch of complex tones can be predicted from the time delay at the first maximum peak of the ACF, τ_1. This result confirmed that obtained by Yost,[2] who demonstrated that pitch perception of iterated rippled noise is greatly affected by the first major ACF peak of the stimulus signal.

Frequency Limits of the Autocorrelation Function Model for Pitch Percept

For fundamental frequencies of 500, 1000, 1200, 1600, 2000, and 3000 Hz, stimuli consisting of two or three pure tone components were produced.[3] The two-component stimuli consisted of the second and third harmonics of the fundamental frequency, and the three-component stimuli consisted of the second, third, and fourth harmonics. The starting phase of all components was adjusted to zero (in phase). The total SPL at the center of the listener's head was fixed at 74 dB. The ACF of all stimuli was calculated obtaining the peak τ_1 related to the fundamental frequency (i.e., $1/\tau_1$). The loudspeaker was placed in front of a subject in an anechoic chamber. The distance between the center of the subject's head and the loudspeaker was 80 cm. Three subjects with musical experience (two male and one female, ages between 21 and 27 years) participated. Pitch-matching tests were conducted using complex tones as test stimuli and a pure tone generated by a sinusoidal generator as a reference.

Results for all subjects are shown in Fig. 7.5. Whenever the fundamental frequency of the stimulus was

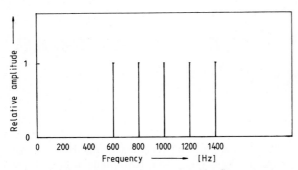

FIG. 7.1 Complex tones tested with five pure-tone components at identical amplitude of 600, 800, 1000, 1200, and 1400 Hz.

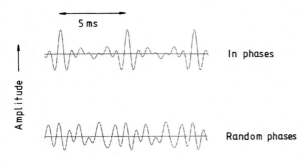

FIG. 7.2 Waveforms of the complex tone with five pure-tone components in phase and random phase.

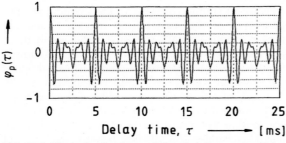

FIG. 7.3 Normalized autocorrelation function analyzed for the complex tone with the five pure-tone components in both phase and random phase. Mathematically it is identical with any phase condition.

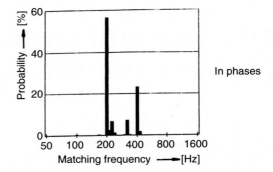

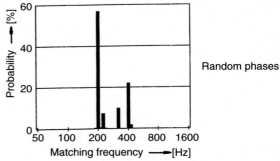

FIG. 7.4 Probability obtained by pitch-matching tests for five subjects adjusted.

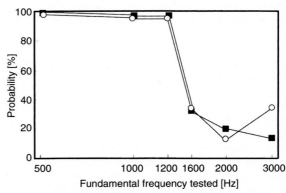

FIG. 7.5 Probability obtained by pitch-matching tests for three subjects adjusted to a pure tone near the fundamental frequency of two complex tones. *Empty circles* represent results for two frequency components, and *full squares* represent those for three components. This shows the limitation up to pitch frequency of approximately 1200 Hz for the autocorrelation function model.

500, 1000, or 1200 Hz, more than 90% of the responses obtained from all subjects under both conditions clustered around the fundamental frequency. However, when the fundamental frequencies of the stimuli were 1600, 2000, or 3000 Hz, the probability that the subjects adjusted the frequency of the pure tone to the calculated fundamental frequency, decreased significantly. These results imply that the ACF model is applicable when the fundamental frequency of stimuli is less than 1200 Hz.

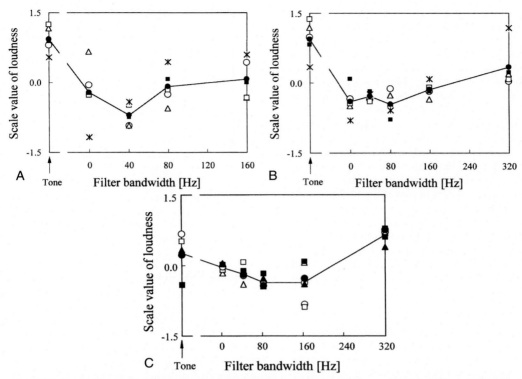

FIG. 7.6 Scale values of loudness obtained by the paired-comparison test as a function of the bandwidth of noise by applying sharp filters with a cut-off slope of 2068 dB/octave. Different symbols indicate the scale values obtained with different subjects. (A) f_c = 250 Hz. (B) f_c = 500 Hz. (C) f_c = 1000 Hz.

Loudness of Sharply Filtered Noise Varying τ_e Within the Critical Band

In this section, we will show that loudness of sharply filtered noise with identical SPL is not constant, albeit within the critical band. Tests revealed that loudness of the pure tone was significantly larger than that of sharply filtered noises, and loudness increased with increasing τ_e extracted from ACF within the critical band.

The bandwidth (Δf) of a sharp filter was changed by using a 2068 dB/octave cut-off slope, which was realized by a combination of two filters.[4,17] Factors of τ_1, τ_e, and ϕ_1 were extracted from the ACF. In fact, the filter bandwidth of 0 Hz included only its slope component. All source signals were the same SPL at 74 dBA, which was accurately adjusted by measurement of the ACF at the origin of the delay time, $\Phi(0)$.

The loudness judgment was performed by the paired-comparison test (PCT) for which the ACF of the band pass noise was changed. A headphone delivered the same sound signal to the two ears. Thus the

interaural cross-correlation (IACC) was kept constant at near unity. Sound signals were digitized at a sampling frequency of 48 kHz. Five subjects with normal hearing were seated in an anechoic chamber and asked to judge which of two paired sound signals they perceived as louder. Stimulus durations were 1.0 seconds, rise and fall times were 50 ms, and silent intervals between the stimuli were 0.5 seconds. A silent interval of 3.0 seconds separated each pair of stimuli, and the pairs were presented in random order.

Fifty responses (5 subjects × 10 sessions) to each stimulus were obtained. Consistency tests indicated that all subjects had a significant ($P < .05$) ability to discriminate loudness. The test of agreement also indicated that there was significant ($P < .05$) agreement among all subjects. A scale value (SV) of loudness was obtained by applying the law of comparative judgment (Thurstone case V) and was confirmed by goodness of fit.

The relationship between the SV of loudness and the filter bandwidth is shown in Fig. 7.6A–C. The SV difference of 1.0 corresponds to approximately 1 dB due to the preliminary experiment. For all three-center frequencies (250, 500, 1000 Hz) the SV of loudness is maximal for the pure tone with the infinite value of τ_e and large bandwidths, with minima at smaller bandwidths (40, 80, 160 Hz, respectively). From the dependence of τ_e on filter bandwidth, we found that loudness increases with increasing τ_e within the "critical bandwidth." Results of analysis of variance for the SVs of loudness indicated that for all center frequencies tested, the SV of loudness of the pure tone was significantly larger than for other band pass noises within the critical band ($P < .01$). Consequently, loudness of the band pass noise with identical SPL was not constant within the critical band. In addition, loudness of the pure tone was significantly larger than for sharply filtered noises, and loudness increased with increasing τ_e within the critical band. Therefore we can consider that a whistle and a yodel produced in a plateau may be heard even at far distances from the sound sources.

Duration Sensation

We investigated the duration sensation (DS), the fourth temporal and primary percept, by the PCT. It is known that duration is indicated in musical notation. Results show that relatively shorter DS is observed for a higher pure tone (3000 Hz) than for 500 Hz. In addition, DS of a 500-Hz pure tone is similar to that of a complex tone with 500-Hz pitch.

Experiments for pure and complex tones were performed by the PCT.[5] Throughout this investigation, the SPL was fixed at 80 dBA. Waveform amplitudes during stimulus onsets and offsets were ramped with rise/fall times of 1 ms for all stimuli tested, the time required to reach a threshold 3 dB below the steady level. Perceived durations of two-component complex tones (3000 and 3500 Hz) having a fundamental at 500 Hz were compared with those evoked by pure tone stimuli at 500 and 3000 Hz. Pairs consisting of two stimuli were presented randomly to obtain SVs for DS. Three signal durations, including rise/fall segments, were used for each of the stimuli: D = 140, 150, and 160 ms. There were thus nine stimulus conditions and 36 pair-wise stimulus combinations. The source stimuli were presented in a darkened soundproof chamber from a single loudspeaker at the horizontal distance of 74 ($\pm$1) cm from the center of the seated listener's head. Ten students (aged between 22 and 36 years) participated in both experiments as subjects with normal hearing level. Each pair of stimuli was presented 5 times randomly during every session for each subject.

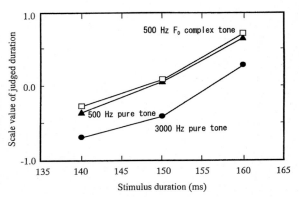

FIG. 7.7 Scale values of duration sensation (DS) obtained by the PCT. $\square$, Complex tone ($F_0 = 500$ Hz) with pure-tone components of 3000 Hz and 3500 Hz; $\blacktriangle$, 500-Hz pure tone; $\bullet$, 3000-Hz pure tone.

Observed SVs for the perceived durations of the nine stimuli are shown in Fig. 7.7. Although signal duration and stimulus periodicity had major effects on perceived duration, the number of frequency components (1 vs. 2) did not. Perceived durations of tones with the same periodicity (f = 500 Hz and F_0 = 500 Hz) were almost identical, whereas durations for pure tones of different frequencies (f = 500 Hz vs. f = 3000 Hz) differed significantly, by approximately 10 ms (judging from equivalent SVs, the 500 Hz pure tone appeared ~10 ms longer than the 3000 Hz tone). Thus the duration (DS) of the higher frequency pure tone (3000 Hz; $\tau_1 = 0.33$ ms) was found to be significantly shorter ($P < .01$) than the duration of either the pure tone (frequency: 500 Hz; $\tau_1 = 2$ ms) or the complex tone (fundamental frequency: $F_0 = 500$ Hz; $\tau_1 = 2$ ms). In addition, the SVs of DS between the two pure tones: $\tau_1 = 2$ (500 Hz) and 0.33 ms (3000 Hz) are almost parallel, so that the effects of periodicity (τ_1) and signal duration (D) on the apparent duration (DS) are independent and additive. Therefore, for these experimental conditions, we may tentatively express that

$$S_L = f(\tau_1, D) = f(\tau_1) + f(D) \qquad (7.1)$$

where τ_1 is extracted from the stimulus ACF. However, we recommend further investigations on apparent duration for a wider range of pitches of complex tone and pure tone.

Timbre of an Electric Guitar Sound With Distortion

Timbre is defined as an aspect of sound quality independent of loudness, pitch, and duration. It is the quality of sound texture that distinguishes two notes of equal pitch, loudness, and duration that are played by different musical instruments. We made an attempt to investigate the relationship between the temporal factors extracted from the ACF of an electric guitar sound and dissimilarity representing the difference of timbre with a difference of distortion.

As shown in Fig. 3.4, a factor $W_{\phi(0)}$ is defined by the delay time at the first 0.5 crossing of the normalized ACF, $\phi(\tau)$, that is deeply related to a global frequency component of the source signal. This value is equivalent to the factor W_{IACC} extracted from the interaural cross-correlation function (IACF).

An electric guitar with "distortion" is a main instrument in pop and rock music. Previously, Marui and Martens[6] investigated timbre variations by use of three types of nonlinear distortion processors with differing level of Zwicker sharpness.[7] Resulting dissimilarity data revealed three dimensions, one of which distinguished between effect types, while the other two were highly correlated with ratings on adjective scales anchored by the pairs dark-bright and sharp-dull. However, no attempt has been made in relation to objective parameters for describing dissimilarity data.

In this study, we examined whether or not timbre can be described by the temporal factors extracted from the running ACF of the source signal.

Experiment A

The purpose of this experiment was to find the factor extracted from the running ACF contributing to the dissimilarity of sounds changing the strength of distortion by the use of a computer. The distortion of music signal $p(t)$ was processed by a computer program, such that: when $|p(t)| \leq C$

$$p(t) = p(t) \qquad (7.2a)$$

and when $|p(t)| \geq C$

$$p(t) = +C, p(t) \geq C; p(t) = -C, p(t) \leq -C \qquad (7.2b)$$

where C is the cut-off pressure amplitude, and its level is defined by

$$CL = 20\log_{10}\left(C/|p(t)|_{max}\right) \qquad (7.3)$$

and $|p(t)|_{max}$ is the maximum amplitude of the signal.

The value of CL was varied between 0 and −49 dB (7 dB step), so that eight stimuli were applied for test signals. As indicated in Table 7.1, pitch, signal duration, and listening level (LL) were fixed. Nineteen student subjects participated (male and female, 20 years of age) who listened to three stimuli and judged dissimilarity. The number of combinations of this experiment was $_8C_3 = 56$ triads. We compiled a dissimilarity matrix according to the judgments, marking the most different pair by "2," the neutral pair by "1," and the most similar pair by "0." After

TABLE 7.1
Conditions of Two Experiments

Condition	Experiment 1	Experiment 2
1) Conditions fixed		
Note (Pitch)	A4 (220 Hz)	A4 (220 Hz)
	By use of 3rd string and 2nd fret	By use of 3rd string and 2nd fret
Listening level in L_{AE} [dB]	80	70
Signal duration [s]	4.0	1.5
2) Conditions varied		
CL [dB] by Equation (7.3)	Eight signals tested changing the cut-off level for 0-49 dB (7 dB step)	
Distortion type	-	Three different types: VINT, CRUNCH and HARD
Drive level	-	Three levels due to the strength of distortion: 50, 70, 90 due to effectors Type ME-30 (BOSE)

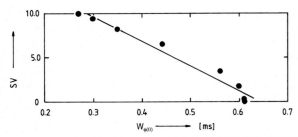

FIG. 7.8 Results for the scale value obtained by regression analysis as a function of the mean value of $W_{\phi(0)}$ (Experiment A).

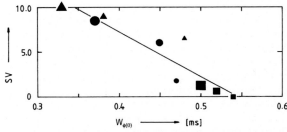

FIG. 7.9 Results for the scale value obtained by regression analysis as a function of the mean value of $W_{\phi(0)}$ (Experiment B).

the analysis of multidimensional scaling, we obtained the SV.

We analyzed contributions to the SV of factors including the mean value of $W_{\phi(0)}$, the decay rate of SPL/s [dBA/s], and the mean value of ϕ_1 (pitch strength). We found that the most significant factor contributing to the SV was the mean value of $W_{\phi(0)}$. Certain correlations between the mean value of $W_{\phi(0)}$ and other factors were found, but the mean value of $W_{\phi(0)}$ is considered as a representative factor. The SV as a function of the mean value of $W_{\phi(0)}$ is shown in Fig. 7.8. The correlation between the SV and the value of $W_{\phi(0)}$ was the most significant: 0.98 ($P < .01$).

Experiment B

Using commercial effectors, we conducted this experiment to find the factor extracted from the running ACF contributing to the dissimilarity of sounds by changing the strength of distortion. As indicated in Table 7.1, we produced nine stimuli with three kinds of effect types (VINT, CRUNCH, and HARD) and three drive levels due to strength of distortion (i.e., 50, 70, and 90 by the effectors Type ME-30 [BOSE]). Twenty student subjects participated (male and female, 20 years of age). The method of experiment was the same as mentioned earlier. Thus the number of combinations of this experiment was $_9C_3 = 84$ triads. Results achieved were similar to Experiment 1, as shown in Fig. 7.9. The correlation coefficient between the scale value (SV) and the value of $W_{\phi(0)}$ was 0.92 ($P < .01$).

We found a common result from the two experiments that indicated the most effective factor in timbre or dissimilarity judgments extracted from the running ACF of the source signal to be $W_{\phi(0)}$.

Concluding Remarks

It is obvious that the previous four primary percepts are all described by factors extracted from the running ACF as shown in the upper part of Fig. 7.10. By contrast, it

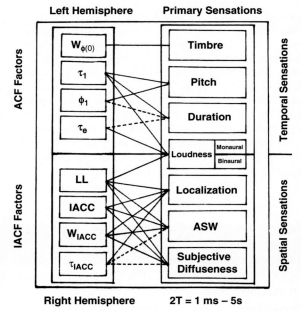

FIG. 7.10 Summarization of temporal and spatial primary percepts in relation to the temporal factors extracted from autocorrelation function and the spatial factors extracted from interaural cross-correlation function, respectively. *ACF*, Autocorrelation function; *ASW*, apparent source width; *IACF*, interaural cross-correlation function; *LL*, listening level.

has been difficult to apply any parameter from the spectral analysis of a sound signal for this purpose. It is worth noting that the single factor $W_{\phi(0)}$ is deeply related to a global frequency component of the source signal. It is noteworthy that Ohgushi[8] showed the lowest and highest frequency components to primarily govern timbre. Remarkably, these two components may be replaced by the single factor $W_{\phi(0)}$.

SPATIAL PERCEPTS IN RELATION TO SPATIAL FACTORS OF THE SOUND FIELD

Localization of a Sound Source in the Horizontal and Median Plane

Localization of a sound source is the most basic spatial percept. Let us now discuss localization in relation to the spatial factors that can be extracted from the IACF.

Localization of a sound source in the horizontal plane may be described essentially by the binaural spatial factors, such that

$$L_{horizontal\ plane} = f[\Phi_{ll}(0), \Phi_{rr}(0), \tau_{IACC}, IACC, W_{IACC}] \quad (7.4)$$

where $\Phi_{ll}(0)$ and $\Phi_{rr}(0)$ are the sound energies arriving at two ear entrances, τ_{IACC}, IACC and W_{IACC} are defined in Fig. 4.2. Movement of a sound source on stage may be described by these spatial factors as a function of time.

A significant spatial factor for determining subjective diffuseness of the sound field is the IACC. A sharp localization percept agrees with a high value of IACC and a narrow value of W_{IACC}.

Cues of Localization in the Median Plane

Five spatial factors for localization in the horizontal plane are given by Eq. (7.4). Of particular interest is the localization in the median plane. These spatial factors extracted from the IACF are not significantly changed as a function of elevation (i.e., due to an almost symmetrical shape of the head and pinnae). This means that IACC remains almost at unity, τ_{IACC} ~ zero, and $\Phi_{ll}(0)$ ~ $\Phi_{rr}(0)$ ~ constant, and W_{IACC} ~ constant, which depend only on the spectrum of the source signal. Therefore cues must be found in the spectrum of the sound signals; however, it is hard to find distinct cues in the head-related transfer function (HRTF) in the spectrum. It is an interesting fact that the temporal factors extracted from the early delay range of the ACF of a sound signal arriving at the ear entrances may act as cues, as discussed later.[9]

We can clearly see differences in the three temporal factors, τ_e, τ_1, and ϕ_1, extracted from the ACF and those calculated by the head related transfer function (HRTF) as a function of the incident angle. According to the model described in Chapter 2 of section "Central Auditory Signal Processing Model," sound localization in the median plane is assumed by a process associated with "left hemisphere specialization" because the related factors may be extracted from the ACF, such that

$$L_{Median} = S_L = f_L(\tau_e, \tau_1, f_1) \quad (7.5)$$

Therefore, as far as monaural localization in the median plane is concerned, we need a learning process

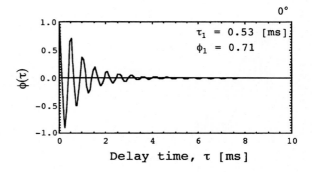

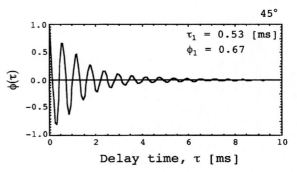

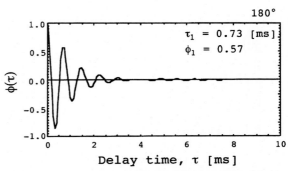

FIG. 7.11 Examples of autocorrelation function for different incident angles in the median plane, 0 degree, 45 degrees, and 180 degrees.

similar to recognition of speech (cognition to recognition). The amplitudes of the transfer functions for a sound incident from the median plane to the ear entrances as measured by Mehrgardt and Mellert,[10] were transformed into the ACF. The following steps produced the ACF:

1. Data sets were obtained from the figures from the paper by Mehrgardt and Mellert using an optical image reader (scanner), with 300 data points each.
2. The amplitude as a function of frequency in logarithmic scale was obtained.

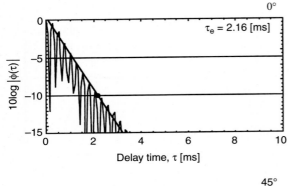

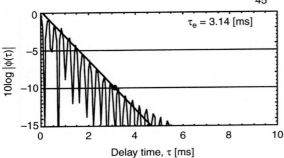

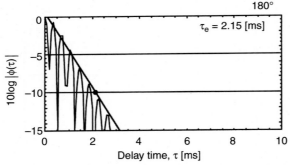

FIG. 7.12 Examples of effective duration extracted from the autocorrelation function envelope in logarithm for different incident angles in the median plane, 0 degree, 45 degrees, and 180 degrees.

3. The amplitude in decibel scale was converted to a real number, and the ACF was calculated by the inverse Fourier transform after passing through the A-weighting filter.

Examples of the normalized autocorrelation function (NACF) are shown in Fig. 7.11. There is a certain degree of correlation between both τ_n and τ_{n+i} and ϕ_n and ϕ_{n+i}, where τ_n and ϕ_n are the delay time and amplitude, respectively, of the n-th peak of the NACF. Thus τ_1,

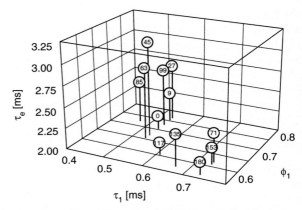

FIG. 7.13 Three-dimensional illustration plotted for three factors, ϕ_1, τ_1, and τ_e, distinguishing different incident angles in the median plane. The number in the circles indicates the incident angle in degrees in the median plane.

and ϕ_1 can be representatives for sets of τ_n and ϕ_n. Examples of plotting the amplitude of the ACF in logarithmic scale as a function of delay time are shown in Fig. 7.12. A straight line can fit the envelope of the decay of the ACF in logarithmic scale, and τ_e was easily obtained from the delay at which the envelope drops to less than −10 dB. The value $\Phi(0)$ is not considered as a cue for sound localization here. Three factors extracted from the NACF are shown in Fig. 7.13. The value of τ_1 for incident angles from 0 to 45 degrees is almost the same, but τ_e for incident angle 45 degrees ($\tau_e = 3.1$ ms) is much larger than at 0 degree ($\tau_e = 2.1$ ms). The value of τ_1 for an incident angle of 180 degrees is different than for the previous two angles; however, ϕ_1 is relatively small.

Obviously, the angle in the median plane can be distinguished by the three factors τ_1, ϕ_1, and τ_e. These factors may therefore play an important role in the perception of localization in the median plane. However, we must consider that certain effects in the learning process of speech recognition are necessary for localization in the median plane as well.

Learning Effects on Monaural Localization in Normal Hearing Listeners
Experimental procedure
We tested localization abilities with and without binaural effects in the free-field median plane.[11] Seven normal-hearing subjects (students 20 to 22 years old) were tested under two test conditions (i.e., binaural, and monaural with the right ear plugged). Each subject

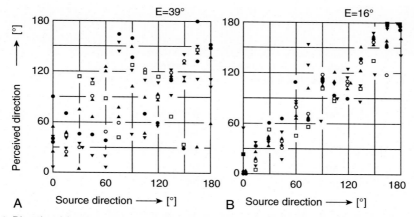

FIG. 7.14 Directional hearing diagram in the median plane of seven normal subjects with the right ear occluded. Different symbols indicate results of localization by different subjects. (A) Without any training, n = 0. (B) After 15 times of training, n = 15.

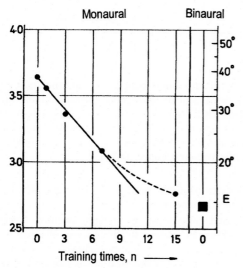

FIG. 7.15 Decrease of localization errors averaged (arithmetic mean) with seven subjects according to training times, n. Solid line: calculated values by Eq. (7.6) with constants $E_0 = 39$ degrees and $\alpha_a = 0.08$. ●, Monaural condition of seven normal subjects with the right ear occluded, n = 0–15; ■, binaural condition of seven subjects without any training, n = 0.

was set at the center of an anechoic chamber, and the head was fixed to avoid movement effects. Thirteen loudspeakers were placed every 15 degrees in the median plane and presented a band pass noise of 106 to 4600 Hz (1.0 seconds), 48 dB(A), switching loudspeakers in random order. Loudspeaker frequency characteristics for 100 to 10,000 Hz were almost flat within 2 dB.

First, the experiment was conducted with normal subjects in binaural listening conditions without any forced learning (n = 0). For every test the subject responded by marking perceived direction on a semicircle printed on a sheet of paper.

Next, the two ears of subjects were plugged with clayey soft rubber and ear protectors, which shifted the threshold level. This way, we ensured that the SPL at 48 dB(A) could not be heard perfectly by the subjects. With the right ear occluded, the localization test was conducted in the monaural condition. Following every presentation in random order from 13 loudspeakers, a correct direction was informed from another loudspeaker placed below the chair the subject was sitting on. In this manner, the localization test and learning were repeated. The first session included no learning (n = 0), the second session included one-time learning (n = 1), the third session included three-time learning (n = 3), the fourth n = 7, and the fifth n = 15. The total time needed for these sessions was approximately 25 minutes.

Experimental results

Fig. 7.14A and B show response results from seven subjects with the right ear occluded in the first session (n = 0) and the fifth session (n = 15), respectively. Localization results of the first session scattered and the averaged error was large, E = 39 degrees; however,

TABLE 7.2
Hearing Levels of 13 Monaural Subjects, Who Had Normal Hearing in One Ear and Congenital Deafness in the Other

Subject	Age (yrs.)	Hearing Loss at 2 kHz of the Left Ear [dB]	Hearing Loss at 2 kHz of the Right Ear [dB]	Remarks
1F[1]	6	80	−5	congenital deafness
2F	7	> 90	0	congenital deafness
3M	9	−5	> 90	congenital deafness
4F	9	0	75	since just after birth
5M	10	> 90	0	congenital deafness
6F	12	−5	85	since 5 months old
7F	13	> 90	−5	since 5 years old
8M	16	75	0	Since 4 years old
9F	18	> 90	0	congenital deafness
10M	20	−5	> 90	congenital deafness
11M	24	50	5	since 1 year old
12M	33	0	85	since childhood
13F	40	0	50	since childhood

[1] F: female, M: male.

the error after learning at n = 15 was dramatically decreased to E = 16 degrees.

Let us now try to formulate the average error, E, as a function of learning time, n.

$$E = E_0 \exp(-\alpha n) \qquad (7.6)$$

where E_0 is the error at n = 0, and αn is a constant expressing decay of error (i.e., an effect of learning).

To examine Eq. (7.6), Fig. 7.15 shows the average error in logarithmic scale and the real number (right vertical scale) obtained against learning time, n.

The solid line from values calculated according to Eq. (7.6) is well fitted with the training number, n < 15. However, at n = 15, a discrepancy is observed with the calculated value, which approaches the error of binaural condition at n = 0.

Discussion
It is interesting to observe that when talking to a newborn baby from a lateral side, he/she reacts to see the person speaking, thereby ensuring the direction of the sound. This is instinctive behavior in binaural localization of a sound source in the horizontal plane performed without any forced learning. Thus the error even at n = 0 for subjects was small enough; the averaged value was approximately E ∼ 14 degrees, as shown on the right side of Fig. 7.15. To attain good localization

ability in the monaural condition with the right ear plugged, normal-hearing listeners need forced learning (cognition and recognition process) more than 15 times. This process is similar to learning languages based on the ACF cues, which may be performed in the left hemisphere (Sperry, 1974).[11a]

Age Effects on Localization in Monaurally Impaired Listeners
As discussed previously, we have examined forced learning effects in relation to the number of times but without any training in this experiment. Here we examined whether or not localization errors can be formulated by Eq. (7.6) in relation to the age of monaurally impaired listeners for sound sources in both the median plane and the horizontal plane.[11]

Experimental procedure
At the university hospital of Kobe University, we selected 13 subjects aged between 6 and 40 years from among patients who had normal hearing in one ear and long-term hearing loss of more than 50 dB at 2 kHz due to otitis media including fever or congenital deafness in the other, as shown in Table 7.2. Typical examples of measured audiograms for three subjects are shown in Fig. 7.16.

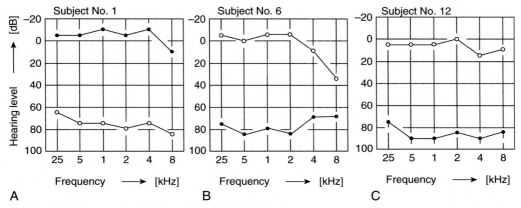

FIG. 7.16 Examples of measured audiograms of monaural subjects. (A) Subject No. 1 (6 years of age). (B) Subject No. 6 (12 years of age). (C) Subject No. 12 (33 years of age). ●, Right ear; ○, left ear.

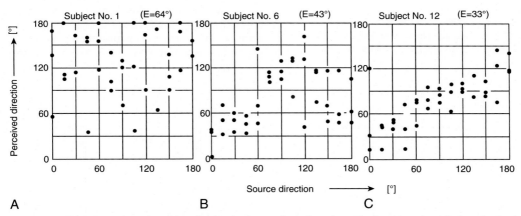

FIG. 7.17 Examples of results of localization in the median plane (n = 0) of monaural subjects, who had normal hearing in one ear and deafness in the other, as indicated in Fig. 7.16. (A) Subject No. 1 (6 years of age). (B) Subject No. 6 (12 years of age). (C) Subject No. 12 (33 years of age).

Band-limited white noise of 1 second (106 to 9600 Hz) was presented from 13 source directions, and sessions were repeated 3 times. The SPLs were 48 to 60 dB(A) which ensured monaural conditions similar to those previously mentioned. Heads were fixed, avoiding effects of movement. Subjects younger than 10 years of age were carefully instructed by an investigator before and during tests.

Experimental results

Examples of experimental results in the median plane of subjects No. 1 (6 years old), No. 8 (16 years old), and No. 12 (33 years old) are shown in Fig. 7.17A–C, respectively. Localization errors averaged for these

subjects were apparently decreased according to age, so that E = 64, E = 45, and E = 35, respectively.

To examine Eq. (7.6) as a function of the subjects' age, results of the averaged error are shown in Fig. 7.18 (r = −0.79, P < .01) for the median plane, and Fig. 7.19A (r = −0.93, P < .01) and Fig. 7.19B (r = −0.87, P < .01) for the horizontal plane of normal ear side and deaf ear side, respectively. According to the fact that for all conditions the averaged localization may be fitted well by straight lines in logarithmic scale (i.e., exponential decays), Eq. (7.6) held to calculate the error as a function of age for both the median and horizontal planes. In addition, results of localization in the horizontal plane were essentially similar for both ear

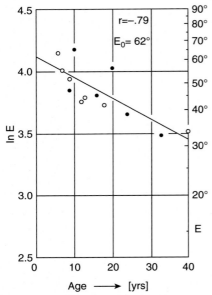

FIG. 7.18 Errors of median plane localization of congenital monaural subjects against age (13 subjects). ●, Male; ○, female.

sides, and average values of coefficient α in the equation were similar to approximately 0.040, which were much larger than for the median plane (i.e., 0.016). This showed that long-time learning effects due to age are much more impactful on localization in the horizontal plane than in the median plane.

Discussion

Age effects on localization ability of monaurally impaired listeners in the horizontal plane were much greater than in the median plane (see Figs. 7.18 and 7.19). This fact is similar to normal-hearing listeners, which may be explained by the condition that daily learning takes place in the horizontal plane more often than in the median plane. It is a well-known fact that in the case of normal-hearing listeners, binaural cues as discussed in Chapter 4 of section "Spatial Factors of the Sound Field" play an important role in localization.

Age effects on localization error of monaurally impaired listeners in the median plane decreased to approximately 32 degrees (see Fig. 7.18). Averaged error in the horizontal planes decreased by the age of monaurally impaired listeners to approximately 15 degrees at 40 years of age. According to the

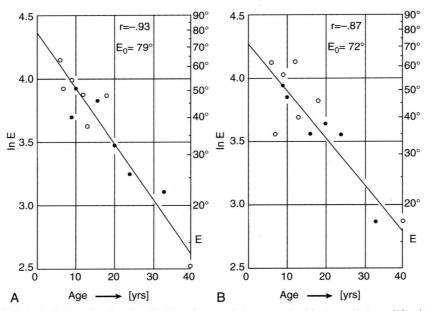

FIG. 7.19 Errors of horizontal plane localization of congenital monaural subjects against age (13 subjects). (A) Normal ear side. (B) Deaf ear side. ●: Male, ○: female.

extrapolation of the calculated value, a remarkable potential is that the error in the median plane may decrease to 15 degrees at the same level (see Fig. 7.19) of normal binaural subjects after 40 years old, say, at 90 years old (extrapolated).

Conclusions

Concluding remarks obtained by two experiments are listed as follows.

1. When we investigated forced learning effects on monaural localization in the median plane for normal subjects (students 20 to 22 years old), we found that the error decreased up to 15 times and approached that of binaural localization. The three monaural cues are illustrated in Fig. 7.13.
2. Both effects of forced learning and natural learning in localization abilities or averaged localization errors as a function of time (number of training sessions and age) decreased exponentially (i.e. straight lines in logarithmic scales) ($P < .01$; see Figs. 7.15 and 7.19).
3. Remarkably, effects of age on averaged error decreased by age in monaurally impaired listeners in both horizontal planes by approximately 15 degrees at 40 years of age (see Fig. 7.19), which is just the same as the level of averaged binaural error in normal listeners 20 to 22 years of age (see Fig. 7.18).
4. Localization errors in the horizontal planes of the deaf ear side and the normal ear side in monaurally impaired listeners do not differ fundamentally, as shown in Fig. 7.19A and B.

5. Age effects on localization ability of monaurally impaired listeners in the horizontal plane were much greater than in the median plane (see Figs. 7.18 and 7.19).

Apparent Source Width for Normal Subjects

Apparent source width (ASW) is one of the spatial percepts for a sound source which is related to spatial factors extracted from the IACF: W_{IACC}, IACC, and LL.

Experiment: apparent source width in relation to W_{IACC} and IACC

Controlling values of W_{IACC} and IACC, the SV of ASW was obtained by the PCT with 10 subjects.[12] To control W_{IACC}, the center frequency of $^1/_3$ -octave band pass noises alternated between 250 Hz, 500 Hz, 1 kHz, and 2 kHz. The values of IACC were adjusted by the sound pressure ratio between reflections ($\xi = \pm 54$ degrees) and the direct sound ($\xi = 0$ degree). To avoid effects of the LL on ASW,[13] in this investigation the total SPL at the ear canal entrances of all sound fields was kept constant at a peak of 75 dBA. Listeners judged which of two sound sources they perceived as wider.

Results of the analysis of variance for the SVs S(ASW) indicate that both factors IACC and W_{IACC} are significant ($P < .01$) and contribute to S(ASW) independently, which yields

$$S_R(ASW) = f(IACC) + f(W_{IACC}) \approx \alpha(IACC)^{3/2} + \beta(W_{IACC})^{1/2}$$

$$(7.7)$$

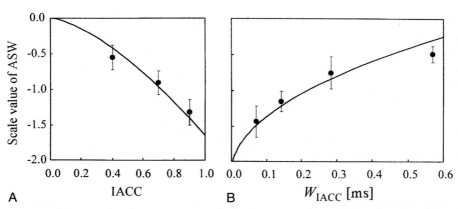

FIG. 7.20 Scale values of apparent source width for $^1/_3$ -octave band pass noises with 95% reliability as a function of (A) the IACC and (B) the W_{IACC}.

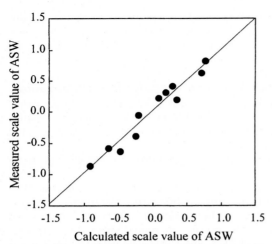

FIG. 7.21 Relationship between measured scale values of apparent source width (ASW) and the scale values of ASW calculated by Eq. (7.7) with α = 1.64 and β = 2.44. Correlation coefficient r = 0.97 (P < .01).

where coefficients $\alpha \approx -1.64$ and $\beta \approx 2.44$ are obtained by regressions of the SVs with 10 subjects, as shown in Fig. 7.20. This holds under the conditions of a constant LL and $\tau_{IACC} = 0$. Obviously, as shown in Fig. 7.21, SVs calculated by Eq. (7.7) and measured SVs are in good agreement (r = 0.97, P < .01).

Experiment: apparent source width in relation to W_{IACC} and listening level

Controlling values of W_{IACC} and LL, the SV of ASW was obtained by the PCT with five subjects.[14] To control W_{IACC}, we applied a complex noise with different frequency components similar to above. To find the effects of LL on ASW, the SPL at the listener's head position was changed from 70 to 75 dB. The values of the IACC of all sound fields were fixed to 0.90 ± 0.01 by controlling the sound pressure ratio of the reflections relative to the level of the direct sound.

Results of the analysis of variance for SVs of the ASW revealed that the explanatory factors W_{IACC} and LL are significant ($P < .01$) (Fig. 7.22). The interaction between W_{IACC} and LL is insignificant, so that we obtain

$$S_R(ASW) = S_R = f(W_{IACC}) + f(LL) \approx a(W_{IACC})^{1/2} + b(LL)^{3/2}$$
$$(7.8)$$

where coefficients a = 2.40 and b = 0.005 were obtained by multiple regression analyses. It is noteworthy that the SV of ASW for the $^{1}/_{3}$ -octave band pass noise is also expressed in terms of the ½ power of W_{IACC}, as expressed by Eq. (7.8), and that the coefficient for W_{IACC} ($\beta \approx 2.44$) is close to that of this study.

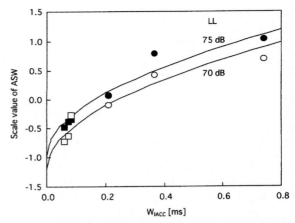

FIG. 7.22 Average scale values of apparent source width as a function of W_{IACC} and as a parameter of listening level (LL). ●, band pass noise; LL = 75 dB; ○, band pass noise; LL = 70 dB; ■, complex noise; LL = 75 dB; □, complex noise; LL = 70 dB. The regression curve is expressed by Eq. (7.8) with a = 2.40 and b = 0.005. *ASW*, Apparent source width.

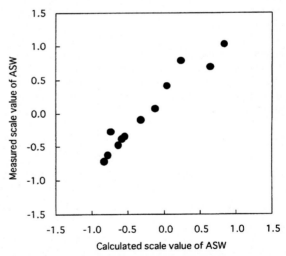

FIG. 7.23 Relationship between the measured scale values of apparent source width (ASW) and the scale values of ASW calculated by Eq. (7.8) with a = 2.40, b = 0.005 and c = −1.60. The correlation coefficient, r = 0.97 (P < .01).

A remarkable result is that the factor W_{IACC} was determined by the frequency component of the source signal, thus the pitch or the fundamental frequency did not influence the ASW.

Fig. 7.23 shows the relationship between the measured SV of ASW and the SV of ASW calculated by

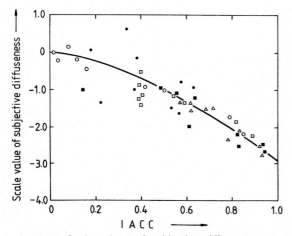

FIG. 7.24 Scale values of subjective diffuseness as a function of IACC (calculated). Different symbols indicate different frequencies of the $\frac{1}{3}$ -octave band pass noise: $\triangle$, 250 Hz; $\bigcirc$, 500 Hz; $\square$, I kHz; $\bullet$, 2 kHz; $\blacksquare$, 4 kHz. (___), Regression line by Eq. (7.10).

Eq. (7.8). The correlation coefficient between the measured and calculated SVs is 0.97 ($P < .01$).

Apparent source width in relation to all three factors, W_{IACC}, listening level, and IACC

It is interesting that the weighting coefficients of $(W_{IACC})^{\frac{1}{2}}$ in two different experimental results are apparently similar, approximately 2.40. Combining Eqs. (7.7) and (7.8) into one yields a single formula

$$S(ASW) = S_R$$
$$= f(W_{IACC}) + f(LL) + f(IACC) \approx a(W_{IACC})^{1/2}$$
$$+ b(LL)^{3/2} + c(IACC)^{3/2} \qquad (7.9)$$

where coefficients are a $\approx$ 2.40, b $\approx$ 0.005, and c $\approx$ −1.60.

At the design stage of an opera house, for example, the SV of ASW may be calculated after getting the spatial factors extracted from IACF at each seating position by the use of an architectural scheme.

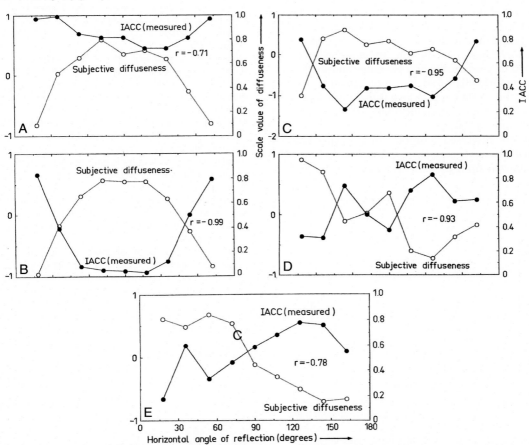

FIG. 7.25 Scale values of subjective diffuseness and IACC as a function of the horizontal angle of incidence to a listener, with $\frac{1}{3}$ -octave band noise of center frequencies. (A) 250 Hz. (B) 500 Hz. (C) 1 kHz. (D) 2 kHz. (E) 4 kHz.

Subjective Diffuseness

To obtain the SV of subjective diffuseness, we conducted the PCT with band pass Gaussian noise, varying the horizontal angle of two symmetric reflections.[15,16] Listeners judged which of two sound fields were perceived as more diffuse. A remarkable finding is that the SVs of subjective diffuseness are inversely proportional to the IACC, and this relationship may be formulated in terms of the $3/2$ power of the IACC in a manner similar to subjective preference values, that is,

$$S \approx -\alpha(\text{IACC})^\beta \qquad (7.10)$$

where $\alpha = 2.9$ and $\beta = 3/2$.

PCT results of the SVs of subjective diffuseness as a function of the IACC calculated by Eq. (7.10) are shown in Fig. 7.24. There is great variation in data in the range of IACC < 0.5; however, no essential difference may be found in the results with frequencies between 250 Hz and 4 kHz. The SVs of subjective diffuseness depending on the horizontal angles are shown in Figs. 7.25 and 7.26 for $1/3$ -octave band pass noise with center frequencies of 250 Hz, 500 Hz, 1 kHz, 2 kHz, and 4 kHz. Clearly, the most effective horizontal angles of reflections depend on the frequency range. For example, these are approximately ±90 degrees for the 500-Hz range and the frequency range less than 500 Hz, approximately ±55 degrees for the 1-kHz range (which is the most important angle for music), and smaller than ±20 degrees for the 2- and 4-kHz range.

So far, we discussed temporal primary percepts and spatial primary percepts in relation to temporal factors extracted from the ACF of sound signals and spatial factors from the IACF, respectively, as summarized in Fig. 7.10.

REFERENCES

1. Sumioka T, Ando Y. On the pitch identification of the complex tone by the autocorrelation function (ACF) model. *J Acoust Soc Am.* 1996;100(A):2720.
2. Yost WA. A time domain description for the pitch strength of iterated rippled noise. *J Acoust Soc Am.* 1996;99:1066–1078.
3. Inoue M, Ando Y, Taguti T. The frequency range applicable to pitch identification based upon the auto-correlation function model. *J Sound Vib.* 2001;241:105–116.
4. Sato S, Kitamura T, Ando Y. Loudness of sharply (2068 dB/Octave) filtered noises in relation to the factors extracted from the autocorrelation function. *J Sound Vib.* 2002;250:47–52.
5. Saifuddin K, Matsushima T, Ando Y. Duration sensation when listening to pure tone and complex tone. *J Temporal Des Arch Environ.* 2002;2:42–47. http://www.jtdweb.org/journal/.
6. Marui A, Martens WL. Constructing individual and group timbre space for sharpness-matched distorted guitar timbres. In: *Audio Engineering Society Convention Paper.* Presented at the 119th Convention New York, 2005.
7. Zwicker E, Fastl H. *Psychoacoustics.* New York: Springer-Verlag; 1999.
8. Ohgushi K. Physical and psychological factors governing timbre of complex tones. *J Acoust Soc Jpn.* 1980;36:253–259.
9. Sato S, Ando Y, Mellert V. Cues for localization in the median plane extracted from the autocorrelation function. *J Sound Vib.* 2001;241:53–56.
10. Mehrgardt S, Mellert V. Transformation characteristics of the external human ear. *J Acoust Soc Am.* 1977;61:1567–1576.
11. Ando Y, Morimoto M, Yorifuji Y, Hattori H. Effects of trainings on accuracy of monaural sound localization. In: *Technical Committee of Hearing Research, the Acoustical Society of Japan.* 1977.
11a. Sperry RW. Lateral specialization in the surgically separated hemispheres. In: Worden FC, Schmitt FO, eds. *The Neurosciences: Third study program.* Cambridge: MIT Press; 1974; Chapter 1.
12. Sato S, Ando Y. Effects of interaural cross-correlation function on subjective attributes. *J Acoust Soc Am.* 1996;100(A):2592.
13. Keet MV. The influence of early lateral reflections on the spatial impression. In: *Proceedings of the 6th International Congress on Acoustics.* Tokyo: Paper E; 1968:2–4.
14. Sato S, Ando Y. Apparent source width (ASW) of complex noises in relation to the interaural cross-correlation function. *J Temporal Des Arch Environ.* 2002;2:29–32. http://www.jtdweb.org/journal/.
15. Ando Y, Kurihara Y. Nonlinear response in evaluating the subjective diffuseness of sound field. *J Acoust Soc Am.* 1986; 80:833–836.
16. Singh PK, Ando Y, Kurihara Y. Individual subjective diffuseness responses of filtered noise sound fields. *Acustica.* 1994;80:471–477.
17. Ando Y. *Auditory and Visual Sensations.* New York: Springer-Verlarg; 2009.

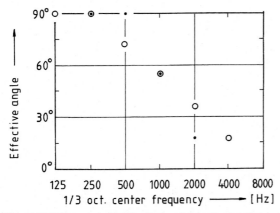

FIG. 7.26 The most effective horizontal angles to a listener that decrease IACC for each frequency band, obtaining the maximum scale value of subjective diffuseness. ○, Angles obtained by the calculated IACC; ●, angles obtained by the observed IACC.

CHAPTER 8

Subjective Preference of the Sound Field

SOUND FIELDS WITH A SINGLE REFLECTION AND MULTIPLE REFLECTIONS
Preferred Delay Time of a Single Reflection

It is well known that a hearing aid system that allows only amplification of a sound signal without the option of adding any reflections subjects the listener to feeling tired even after a short use period because of the psychophysiological effects created by such an unnatural acoustic environment. To begin investigations on the simplest sound field with a single reflection, we obtained basic knowledge toward establishing a theory of subjective preference of the sound field with early multiple reflections and subsequent reverberation of enclosures.[1,2]

The sound field we studied consisted of a frontal direct sound $\xi_1 = 0°(\eta_1 = 0°)$ and a single reflection from a fixed side direction $\xi_1 = 36°(\eta_1 = 9°)$. These angles were selected because they are typical in a room. The delay time Δt_1 of the single reflection arriving after the direct sound was adjusted in the range of 6 to 256 ms. Paired-comparison tests (PCTs) were performed with PhD students and assistants at the Third Physics Institute in Goettingen and Graduate School of Science and Technology, Kobe University, with two different music motifs (A and B) and a speech signal (Table 8.1). The score here was obtained by accumulation, giving $+1$ and -1 to positive and negative judgments, respectively, and dividing the total score by $S(F - 1)$ to get the normalized score, where S is the number of subjects and F the number of sound fields. The normalized scores and the percentage of preference of the sound fields as a function of the delay are shown in Fig. 8.1A and B.[3,4]

Obviously, the most preferred delay time with the maximum score differs greatly between the two motifs and the speech signal. When the amplitude of reflection is $A_1 = 1$, the most preferred delays are approximately 130 ms for music motif A, 35 ms for music motif B (see Fig. 8.1A), and 16 ms for speech (see Fig. 8.1B). We found that these correspond well with the effective duration of the auto correction function (ACF), $(\tau_e)_{min}$,

of source signals of 125 ms (motif A), 40 ms (motif B), and 10 ms (speech), as indicated in Table 8.1. After inspection, the preferred delay was found approximately at a certain duration of ACF, defined by τ_p, such that the envelope of ACF becomes $0.1A_1$. Thus $\tau_p = (\tau_e)_{min}$ only when $A_1 = 1$. The data collected as a function of duration, τ_p, are shown in Fig. 8.2, where data from a continuous speech signal are also plotted. When the envelope of the ACF is exponential, it is approximately expressed by[1]

$$\tau_p = [\Delta t_1]_p \approx (1 - \log_{10}A_1)(\tau_e)_{min} \tag{8.1}$$

It is worth noting that the amplitude of reflection relative to that of the direct sound should be measured by the most accurate method, for example, the square root of ACF at the origin of delay time, $[\Phi_p(0)]^{1/2}$.

Two reasons can explain why preference decreases for the short delay range of reflection, $0 < \Delta t_1 < \tau_p = [\Delta t_1]_p$ (Fig. 8.3):

1. Tone coloration effects occur because of the interference phenomenon in the coherent time region.
2. Interaural cross-correlation function (IACC) increases when Δt_1 is near 0.

On the other hand, echo disturbance effects can be observed when Δt_1 is greater than τ_p.

TABLE 8.1
Music and Speech Source Signals and Their Minimum Effective Duration of the Running ACF, $(\tau_e)_{min}$

Sound Source	Title	Composer or Writer	$(\tau_e)_{min}$ [ms][1]
Music motif A	Royal Pavane	Orlando Gibbons	125
Music motif B	Sinfonietta, Opus 48; IV movement	Malcolm Arnold	40
Speech S	Poem read by a female	Doppo Kunikida	10

[1] The value of $(\tau_e)_{min}$ is the minimum value extracted from the running ACF, 2T = 2 s, with a running interval of 100 ms.

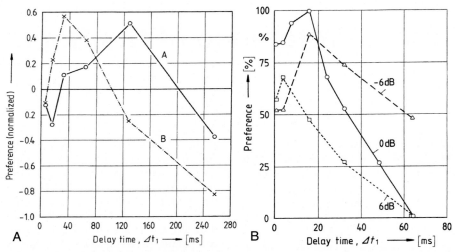

FIG. 8.1 Results of subjective preference for different sound sources as a function of the delay time of a single reflection obtained by the paired comparison test (PCT) giving $+1$ and -1, corresponding to positive and negative judgment, respectively. The normalized score is obtained by the factor $S(F - 1)$, where S is the number of the sound field and F is the number of subjects (6 sound fields and 13 subjects). (A) Normalized preference score for music motif A ($\tau_e \approx 127$ ms) and music motif B ($\tau_e \approx 35$ ms). (B) Percentile preference for a continuous speech signal ($\tau_e \approx 12$ ms). It is worth noting that the most preferred delay time of the reflection is approximately related to the value of τ_e, when $A_1 = 1.0$. ([A], Ando Y. Subjective preference in relation to objective parameters of music sound fields with a single echo. *J Acoust Soc Am.* 1977;62:1436−1441; [B], Ando Y, Kageyama K. Subjective preference of sound with a single early reflection. *Acustica.* 1977;37:111−117.)

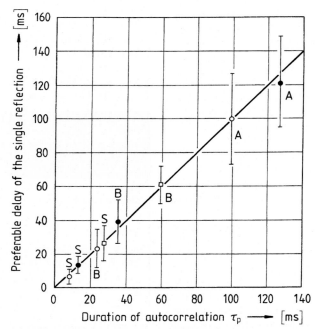

FIG. 8.2 The most preferred delay time of the single reflection as a function of the duration of the autocorrelation function of source signals, τ_p given by Eq. (8.1).

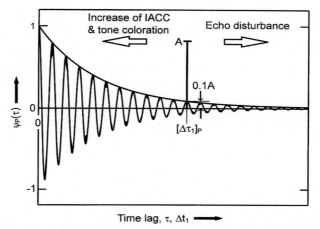

FIG. 8.3 Subjective attributes before and after the most preferred delay time of the reflection, $[\Delta t_1]_p = \tau_p$. *IACC,* Interaural cross-correlation function. (Ando Y. *Concert Hall Acoustics*. Heidelberg: Springer-Verlag; 1985.)

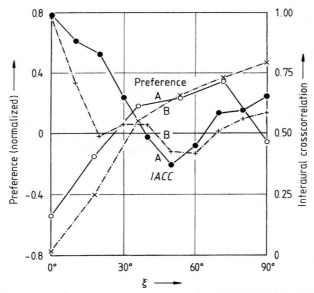

FIG. 8.4 Normalized preference score and the interaural cross-correlation function *(IACC)* for music motif A ($\tau_e \approx 127$ ms) and music motif B ($\tau_e \approx 35$ ms) as a function of the horizontal angle of incidence of the single reflection, ξ. A strong negative correlation between subjective preference and the IACC can be seen. The most effective horizontal angle minimizing the IACC can be found commonly around $\xi \approx 55$ degrees, regardless of sound source or the value of τ_e. (Ando Y. Subjective preference in relation to objective parameters of music sound fields with a single echo. *J Acoust Soc Am*. 1977;62:1436–1441.)

Preferred Horizontal Direction of a Single Reflection to a Listener

Applying music motif A and B, the delay time of the reflection, in the experiment showing the preferred direction of a single reflection, was fixed at 32 ms. The direction was specified by loudspeakers located at

$$\xi_1 = 0°(\eta_1 = 27°), \text{and } \xi_1 = 18°, 36°, ..., 90°(\eta_1 = 9°)$$

Results of the preference tests are shown in Fig. 8.4. No fundamental differences were observed between the curves of the two motifs in spite of the great difference in $(\tau_e)_{min}$. The preferred score increases approximately with decreasing IACC. The correlation coefficient between the score and the IACC is -0.8 ($P < .01$). The score with motif A at $\xi_1 = 90$ degrees drops to a negative value, indicating that the lateral reflections, coming only from approximately $\xi_1 = 90$ degrees, are not always preferred. The figure shows that there is a preference for angles smaller than $\xi_1 = 90$ degrees, and on average there may be an optimum range centered on approximately $\xi_1 = 55$ degrees. Similar results can be seen in the data obtained from speech signal analysis.[4]

SOUND FIELDS WITH EARLY REFLECTIONS AND SUBSEQUENT REVERBERATION

We examined the independence of the effects of the four physical factors on subjective preference judgments. To obtain a four-dimensional continuity of the scale value, two factors were varied simultaneously while the remaining two were held constant. Thus at least three tests must be conducted to cover four orthogonal factors.

The system simulating the sound field in a room is shown in Fig. 4.3. A computer program provided the time delay of two early reflections ($n = 1, 2$) and the subsequent reverberation ($n > 2$), relative to the direct sound. To represent the geometrical size of a similar room shape, the scale of dimension (SD) is introduced as follows:

$$\Delta t_1 = 22(SD), \Delta t_2 = 38(SD), \Delta t_3 = 47(SD) \, [ms] \quad (8.2)$$

The reverberation signal with constant frequency characteristics, which is the preferred condition,[1] was generated by the Schroeder reverberator.[5] To attain a natural sound, the conditions of the simulation system were carefully selected.

Test A: Independence of factors SD and T_{sub} on subjective preference

To examine whether or not the temporal-monaural factors influence the scale values of subjective preference independently, PCTs on the 16 sound fields for each source signal were conducted for changes in SD and T_{sub} with 9 to 14 subjects.[6] Both of the factors are associated with the left hemisphere of the human brain, as mentioned in Chapter 6 of section "Specialization of Cerebral Hemispheres for Temporal and Spatial Factors" (Table 6.1).

Test B: Independence of factors listening level and interaural cross-correlation function on subjective preference

To determine the independent influence of the spatial and right hemispheric factors, listening level (LL) and IACC, PCTs on the 12 sound fields for each source signal were performed with 13 to 14 subjects.[7]

Test C: Independence of factors T_{sub} and interaural cross-correlation function on subjective preference

For reconfirmation of the independence of the left and right hemispheric factors, T_{sub} and IACC, preference tests were conducted for 16 sound fields with eight subjects.[8]

According to results of analyses of variance for the scale values obtained by the law of comparative judgment from three tests, each factor influences the scale value of preference independently.[1]

Other test: Independence of factors scale of dimension and interaural cross-correlation function on subjective preference

In addition, subjective preference judgments performed on sound fields with multiple early reflections[9] and with subsequent reverberation[10] confirmed that factors SD and IACC are independent of each other in the subjective preference judgments.

Thus we may conclude that the total scale value of subjective preference is calculated by the law of superposition in the range of preferred conditions for the four orthogonal factors tested. The consistency of the unit of the scale values obtained from the different preference tests has been discussed at length.[1]

OPTIMAL CONDITIONS MAXIMIZING SUBJECTIVE PREFERENCE

Due to systematic investigations of simulating sound fields in a room by the aid of a computer and the PCT, we are able to describe the design objectives and linear scale value of subjective preference. Optimal design objectives can be described in terms of subjectively preferred sound qualities, which are related to the temporal and spatial factors describing the sound signals arriving at the two ears. They clearly lead to comprehensive criteria consisting of temporal and spatial factors associated with the left and right human cerebral hemispheres, respectively, for achieving optimal design of opera houses, as summarized later.[1,2,11]

Listening Level

The LL is of course the primary criterion for listening to vocal and orchestral music in the sound field of any room. Preferred LL depends on the speech signals and music of a particular passage. For example, the gross preferred levels obtained from 16 subjects are in peak ranges of 77 to 79 dBA for two different musical sound sources (i.e., music motif A [Royal

Pavane by Gibbons] with slow tempo, and 79 to 80 dBA for music motif B [Sinfonietta by Arnold] with fast tempo).

Early Reflections After the Direct Sound (Δt_1)

An approximate relationship for the most preferred delay time $[\Delta t_1]_p$ has been described in terms of the envelope of the autocorrelation function of source signals and the total amplitude of reflections, A.[1] In general, it is expressed by

$$|\phi_p(\tau)| \text{envelope} : kA^c, \text{ at } \tau = [\Delta t_1]_p \qquad (8.3)$$

ϕ where k and c are constants which depend on subjective attributes.[2] When the envelope of the ACF is exponential, as indicated for most of music and speech signals, then

$$[\Delta t_1]_p \approx (\log_{10} 1/k - c\log_{10}A)(\tau_e)_{min} \qquad (8.4)$$

where the total pressure amplitude of reflection is given by

$$A = \left[A_1^2 + A_2^2 + A_3^2 + ...\right]^{1/2} \qquad (8.5)$$

and k = 0.1 and c = 1.

The relationship of Eq. (8.3) for a single reflection may be obtained by entering A = A_1 so that

$$[\Delta t_1]_p \approx (1 - \log_{10}A_1)(\tau_e) \qquad (8.6a)$$

Later experiments have accurately expressed that the value of τ_e in Eq. (8.6a) is replaced by the minimum value of τ_e of the running ACF,[12,13] so that

$$[\Delta t_1]_p \approx (1 - \log_{10}A_1)(\tau_e)_{min} = \tau_p \qquad (8.6b)$$

The value of $(\tau_e)_{min}$ is observed at the most active part of a music piece containing the most "artistic" information like a "vibrato," a "quick" in the music flow, and/or a very sharp sound signal like attack. Echo disturbance may therefore be perceived at the time of attack or a sharp change of a signal in a piece where $(\tau_e)_{min}$ occurs.

Even with long music compositions, the music flow can be divided into such short segments that the minimal part of the whole music that is, $(\tau_e)_{min}$ of the running ACF which determines the preferred temporal conditions, can be used for the choice of music program to be performed in a given room. We have discussed a method of controlling the minimum value $(\tau_e)_{min}$, which determines the preferred temporal conditions for vocal music.[14,15] Introducing vibrato during singing effectively decreases the value of $(\tau_e)_{min}$.

Subsequent Reverberation Time after Early Reflections (T_{sub})

We have observed that the most preferred condition of frequency response to reverberation time is just flat.[12] The preferred subsequent reverberation time is approximately expressed by

$$[T_{sub}]_p \approx 23(\tau_e)_{min} \qquad (8.7)$$

The values A given by Eq. (8.5) tested were 1.1 and 4.1, which cover the usual conditions of sound fields in rooms.

Recommended reverberation times for several sound sources are shown in Fig. 8.5. A lecture and conference

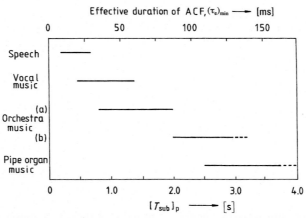

FIG. 8.5 Recommended reverberation time for sound sources including speech that is related to the value of τ_e, that is the effective duration of the normalized ACF, which is defined by the delay τ_e at which the envelope of the normalized ACF becomes 0.1 (−10 dB).

room must be designed for speech, and an opera house for two greatly different sources (i.e., vocal music and orchestral music).

Magnitude of Interaural Cross-correlation Function

All available data from more than 500 listeners indicate a negative correlation between the magnitude of IACC and subjective preference (i.e., humans prefer dissimilarity of signals arriving at the two ears).[16] This holds only under the condition that the maximum value of the IACC is maintained at the origin of time delay, keeping balance of the sound field for the two ears. Otherwise, an image shift of the source may occur. To obtain a small magnitude of IACC in the most effective manner,[2] the directions from which the early reflections arrive at the listener should be kept within a certain range of angles from the median plane (± 55 degrees ± 20 degrees) for usual sound sources consisting of the main frequency range centered on 1.0 kHz. It is obvious that sound arriving from the median plane ± 0 degree makes IACC greater. Sound arriving from ± 90 degrees in the horizontal plane is not always advantageous because similar "detour" paths around the head to both ears cannot decrease the IACC effectively, particularly for frequency ranges higher than 500 Hz. For example, the most effective angles for the frequency ranges of 1 and 2 kHz are approximately centered on ± 55 degrees and ± 36 degrees, respectively.

To realize these conditions simultaneously, we have proposed a geometrical uneven surface.[17]

THEORY OF SUBJECTIVE PREFERENCE FOR THE SOUND FIELD

Let us now put the temporal factors extracted from the running ACF and the spatial factors extracted from the interaural cross-correlation function (IACF), associated with the left and right human cerebral hemispheres, respectively, into a theory of subjective preference.

Because the number of orthogonal acoustic factors included in the sound signals at both ears is limited to four, as mentioned in Chapter 4 of section "Orthogonal Factors of the Sound Field,"[1,2,11] the scale value of any one-dimensional subjective response may be expressed by

$$S = g(x_1, x_2, ..., x_I)$$
$$\equiv g_L + g_R \qquad (8.8)$$

according to the specialization of human cerebral hemispheres (see Chapter 6: Specialization of Cerebral Hemispheres for Temporal and Spatial Factors). In this study, we discuss the linear scale value of preference obtained

by the law of comparative judgment, which is further reconfirmed by procedures.[18–20] We have verified by a series of experiments that four objective factors act independently of the scale value when changing two of four factors simultaneously. Results indicate that the units of scale values are considered to be almost constant, so that we may add scale values to obtain the total scale value,[11]

$$S \equiv g_L + g_R = g(x_1) + g(x_2) + g(x_3) + g(x_4)$$
$$= S_1 + S_2 + S_3 + S_4 \qquad (8.9)$$

where $g_L = S_2 + S_3$ and $g_R = S_1 + S_4$ and S_i, $i = 1, 2, 3, 4$ is the scale value obtained relative to each objective parameter. Eq. (8.9) indicates a four-dimensional continuity. The scale value is relative, and only addition and subtraction operations are allowed.

The dependence of the scale values on each of the four orthogonal factors is shown graphically in Fig. 8.6A through D. From the nature of the scale value, it is convenient to put a zero value at the most preferred conditions, as shown in these figures. The results of the scale value of subjective preference obtained from different test series, using different music programs, yield the following common formula, as indicated by solid lines in Fig. 8.6:

$$S_i \approx -\alpha_i |x_i|^{3/2}, \; i = 1, 2, 3, 4 \qquad (8.10)$$

where x_i is the i-th factor, and the values of α_i are weighting coefficients as listed in Table 6.1. If α_i is close to zero, then a lesser contribution of the factor x_i on subjective preference is signified.

The factor x_1 is given by the sound pressure level difference, measured by the A-weighted network, so that

$$x_1 = 20 \log P - 20 \log[P]_p \qquad (8.11)$$

P and $[P]_p$ are the sound pressure at a specific seat and the most preferred sound pressure that may be assumed at a particular seat position in the room under investigation:

$$x_2 = \log\left(\Delta t_1 / [\Delta t_1]_p\right) \qquad (8.12)$$
$$x_3 = \log\left(T_{sub} / [T_{sub}]_p\right) \qquad (8.13)$$
$$x_4 = IACC \qquad (8.14)$$

The scale values of preference have thus been formulated approximately in terms of the $\{3/2\}$ power of the normalized objective parameters, expressed in logarithm for the parameters, x_1, x_2, and x_3. A remarkable fact is that the spatial binaural parameter x_4 is expressed in terms of the $\{3/2\}$ power of its "real values," indicating a greater contribution to the formation of subjective preference than from other parameters. Thus scale values are not greatly changed in the neighborhood of

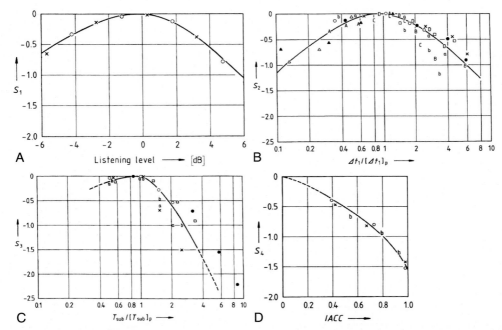

FIG. 8.6 The scale value of subjective preference as a function of each of the four orthogonal factors of the sound field. The maximum scale value is adjusted to zero at the most preferred condition of each factor. (A) Scale value S_1 as a function of listening level. (B) Scale value S_2 as a function of Δt_1 normalized by $[\Delta t_1]_p$ calculated by $[\Delta t_1]_p \sim (1 - \log_{10}A)(\tau_e)_{min}$. (C) Scale value S_3 as a function of T_{sub} normalized by $[T_{sub}]_p$ calculated by Eq. (8.7). (D) Scale value S_3 as a function of interaural cross-correlation function *(IACC)*.

the most preferred conditions but decrease rapidly outside of this range. Because the experiments were conducted to find the optimal conditions, this theory holds in the range of preferred conditions tested for the four factors. This theory has been well based on neural activities in the auditory-brain system that is deeply related to subjective preference for the sound field.[2,21] We have elucidated that two temporal factors—the initial time delay gap between the direct sound and the first reflection, Δt_1, and the subsequent reverberation time, T_{sub}—are associated with the left cerebral hemisphere and that the two spatial factors, the magnitude of IACF, IACC, and the LL are associated with the right.

Applying the theory of subjective preference, we demonstrate the quality of the sound field at each seating position in a concert hall with a shape similar to that of the Symphony Hall in Boston. Suppose that a single source is located at the center, 1.2 m above the stage floor. Receiving points at a height of 1.1 m above the floor level correspond to the ear positions. Reflections with their amplitudes, delay times, and directions of arrival at the listeners are taken into account using the image method.

Contour lines of the total scale value of preference calculated for music motif B are shown in Fig. 8.7. This figure shows effects of the reflections from the side reflectors on stage. The sidewalls on stage may produce decreasing values of IACC for the audience area. Thus the preference value at each seat increases, as shown in Fig. 8.6(B) in comparison with in Fig. 8.6(A). In this calculation, reverberation time is assumed to be 1.8 seconds throughout the hall and the most preferred $[LL]_p = 20\log[P]_p$ in Eq. (8.11) is assumed for a point on the center line 20 m from the source position.

As applications of this theory, the Kirishima International Music Hall[2,22] and the Tsuyama Music Cultural Hall have been designed and built.[23]

SEAT SELECTION ENHANCING INDIVIDUAL PREFERENCE

To maximize individual subjective preference for each listener, a special facility for seat selection, testing each listener's own subjective preference, was first introduced at the Kirishima International Concert Hall in 1994. Sound simulation was realized based on the system shown in

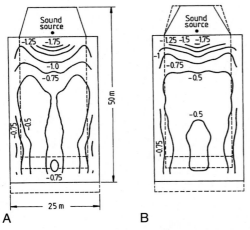

FIG. 8.7 Examples of calculating the total scale value after obtaining the four orthogonal factors at each seating position. (A) An original room shape. (B) Adjusted stage enclosure.

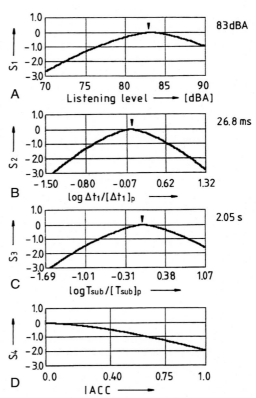

FIG. 8.8 Examples of the scale value of subjective preference obtained by the PCT for each of the four orthogonal factors of the sound field (subject BL). (A) The most preferred listening level was 83 dBA, the individual weighting coefficient in Eq. (8.10): $\alpha_1 = 0.06$. (B) The preferred initial time delay gap between the direct sound and first reflection was 26.8 ms, the individual weighting coefficient in Eq. (8.10): $\alpha_2 = 1.86$, where $[\Delta t_1]_p$ calculated by Eq. (8.10) with $\tau_e = 62$ ms for the music used (A = 4) is 24.8 ms. (C) The preferred subsequent reverberation time was 2.05 seconds, the individual weighting coefficient in Eq. (8.10): $\alpha_3 = 1.46$, where $[T_{sub}]_p$, calculated by Eq. (8.7) with $\tau_e = 62$ ms for the music used, is 1.43 seconds. (D) Individual weighting coefficient in Eq. Eq. (8.10): $\alpha_4 = 1.96$. IACC, Interaural cross-correlation function.

Fig. 4.3 with multiple loudspeakers. The system used arrows for testing subjective preference of sound fields for listeners at the same time. Because the four orthogonal factors of the sound field influence preference judgments almost independently, as we previously discussed, a single orthogonal factor was varied, while the other three were fixed at the most preferred condition for the average listener. Results of testing acousticians who participated in the International Symposium on "Music and Concert Hall Acoustics" (MCHA95), which was held in Kirishima in May 1995, are presented here.[22]

The music source was orchestral, "Water Music" by Handel; effective duration of the ACF, τ_e, was 62 ms. The total number of listeners participating in individual PCTs was 106.[2,24] Typical examples of the test results, as a function of each factor, for listener BL are shown in Fig. 8.8. The scale value of this listener was close to the average collected from previous subjects: the most preferred $[LL]_p$ was 83 dBA, $[\Delta t_1]_p$ was 26.8 ms (the preferred value calculated by Eq. [8.4] was 24.8 ms, where $[\Delta t_1]_p = (1 - \log_{10}A)\,\tau_e$, A = 4), and the most preferred reverberation time was 2.05 seconds (the preferred value calculated by Eq. [8.7] was 1.43 seconds). Thus the center seating area was preferred by listener BL, similarly to the calculated value at the design stage, as shown in Fig. 8.9. With regard to IACC, results from all listeners clearly indicated that the scale value of preference increased with decreasing IACC value. Because listener KH preferred a very short delay time of the initial reflection, his preferred seats

were located close to the boundary walls, as shown in Fig. 8.10. Listener KK indicated a preferred LL exceeding 90 dBA. For this listener, the front seating areas close to the stage were preferable, as shown in Fig. 8.11. On the contrary, for listener DP, whose preferred LL was rather weak (76.0 dBA) and preferred initial delay time short (15.0 ms), the preferred seats were in the rear part of the hall as shown in Fig. 8.12. The preferred initial time delay gap for listener AC exceeded 100.0 ms but

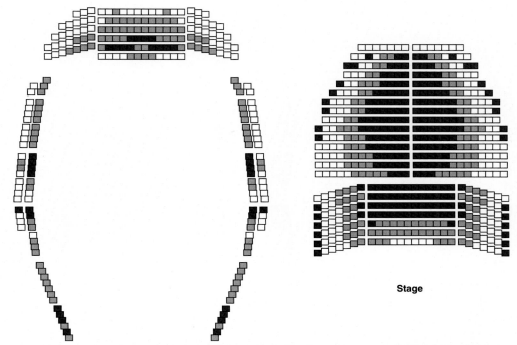

FIG. 8.9 Preferred seating area calculated for subject BL. The seats are categorized into three parts according to the scale value of preference calculated by Eq. (8.9) summating S_1 through S_4. Black portion of the seats indicates preferred areas, approximately one-third of all seats in this concert hall, for subject BL.

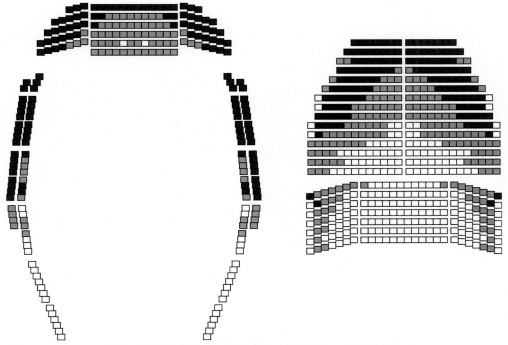

FIG. 8.10 **Preferred seat area calculated for subject KH.**

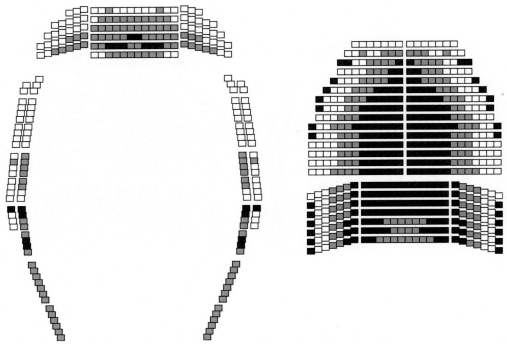

FIG. 8.11 **Preferred seat area calculated for subject KK.**

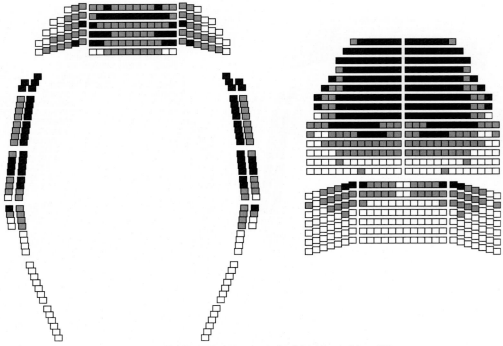

FIG. 8.12 **Preferred seat area calculated for subject DP.**

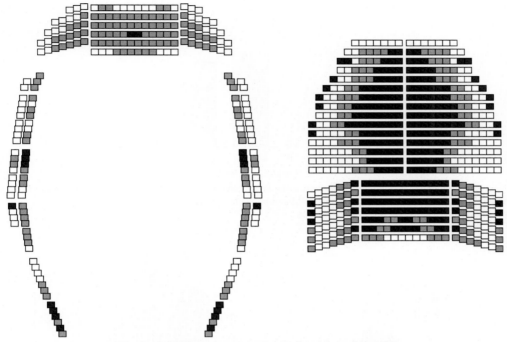

FIG. 8.13 Preferred seat area calculated for subject CA.

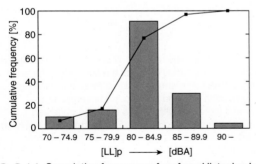

FIG. 8.14 Cumulative frequency of preferred listening level [LL]$_p$ (106 subjects). Approximately 60% of subjects preferred the range of 80 to 84.9 dBA.

was not critical. Thus any initial delay times are acceptable, but IACC is critical. Therefore the preferred area of seats was located only in the rear part, as is shown in Fig. 8.13.

Cumulative frequencies of the preferred value with 106 listeners are shown in Fig. 8.14 through Fig. 8.16 for three factors. As indicated in Fig. 8.14, approximately 60% of listeners preferred the range of 80 to

84.9 dBA in listening to music, but some listeners indicated that the most preferred LL was greater than 90 dBA, and the total range of the preferred LL was scattered, exceeding a 20-dB range. As shown in Fig. 8.15, approximately 45% of the listeners preferred the initial delay times 20 to 39 ms, which were around the calculated preference of 24.8 ms (Eq. 8.4); some listeners indicated 0 to 9 ms and others more than 80 ms. With regard to reverberation time, as shown in Fig. 8.16, approximately 45% of listeners preferred 1.0 to 1.9 seconds which centers on the calculated preferred value of 1.43 seconds, but some listeners indicated preferences less than 0.9 seconds or more than 4.0 seconds.

We assumed that both initial delay time and subsequent reverberation time are related to a kind of "liveliness" of the sound field. In addition, we assumed that a great interference effect on subjective preference arises from between these factors for each individual. However, as shown in Fig. 8.17, there is little correlation between preference values of [Δt_1]$_p$ and [T_{sub}]$_p$ (correlation is 0.06). The same is true for the correlation between values of [T_{sub}]$_p$ and [LL]$_p$ and for that between values of [LL]$_p$ and [Δt_1]$_p$, a correlation of

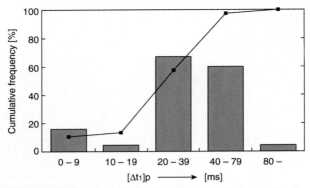

FIG. 8.15 Cumulative frequency of preferred initial time delay gap between the direct sound and the first reflection $[\Delta t_1]_p$ (106 subjects). Approximately 45% of subjects preferred the range of 20 to 39 ms. Calculated value of $[\Delta t_1]_p$ by Eq. (8.4) is 24.8 ms.

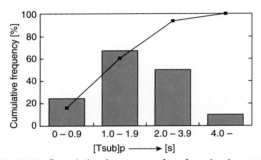

FIG. 8.16 Cumulative frequency of preferred subsequent reverberation time $[T_{sub}]_p$ (106 subjects). Approximately 45% of subjects preferred the range of 1.0 to 1.9 s. Calculated value of $[T_{sub}]_p$ by Eq. (8.7) is 1.43 seconds.

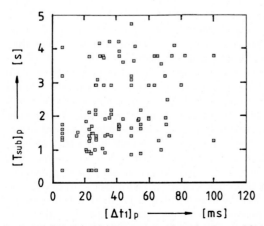

FIG. 8.17 Relationship between preferred values of $[\Delta t_1]_p$ and $[T_{sub}]_p$ for each subject. No significant correlation between them was identified.

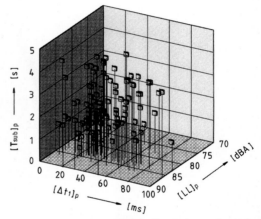

FIG. 8.18 Three-dimensional illustration of preferred factors, $[LL]_p$, $[\Delta t_1]_p$, and $[T_{sub}]_p$, of the sound field for each individual subject. Preferred condition for factor interaural cross-correlation function is excluded because no fundamental individual differences could be observed. Preferred conditions are distributed in a certain range of each factor, so that subjects could not be classified into any specific groups.

less than 0.11. Fig. 8.18 shows the three-dimensional plots of the preferred values of $[LL]_p$, $[\Delta t_1]_p$, and $[T_{sub}]_p$, excluding the consensus factor of IACC. Looking at a continuous distribution in preferred values, no specific groupings of individuals can be classified as emerging from the data.

In calculation with Eq. (8.10), there is no correlation between weighting coefficients α_i and α_j, $i \neq j$, (i and j = 1, 2, 3, 4) either.[2] A listener indicating a relatively small value of one factor will not always indicate a relatively small value for another factor. Thus a listener can be critical about preferred conditions as a function of certain factors, while insensitive to other factors, resulting in individual characteristics distinct from other listeners. This is an indication of individual difference.

It is worth noting that to attain a natural sound field with reflections for a binaural hearing aid, these four orthogonal factors can be used in the design.

REFERENCES

1. Ando Y. *Concert Hall Acoustics*. Heidelberg: Springer-Verlag; 1985.
2. Ando Y. *Architectural Acoustics, Blending Sound Sources, Sound Fields, and Listeners*. New York: AIP Press/Springer-Verlag; 1998.
3. Ando Y. Subjective preference in relation to objective parameters of music sound fields with a single echo. *J Acoust Soc Am.* 1977;62:1436–1441.
4. Ando Y, Kageyama K. Subjective preference of sound with a single early reflection. *Acustica.* 1977;37:111–117.
5. Schroeder MR. Natural sounding artificial reverberation. *J Audio Eng Soc.* 1962;10:219–223.
6. Ando Y, Okura M, Yuasa K. On the preferred reverberation time in auditoriums. *Acustica.* 1982;50:134–141.
7. Ando Y, Morioka K. Effects of the listening level and the magnitude of the interaural cross-correlation (IACC) on subjective preference judgment of sound field. *J Acoust Soc Jpn.* 1981;37:613–618. (in Japanese with English abstract).
8. Ando Y, Otera K, Hamana Y. Experiments on the universality of the most preferred reverberation time for sound fields in auditoriums. *J Acoust Soc Jpn.* 1983;39:89–95. (in Japanese with English abstract).
9. Ando D, Gottlob D. Effects of early multiple reflections on subjective preference judgments on music sound fields. *J Acoust Soc Am.* 1979;65:524–527.
10. Ando Y, Imamura M. Subjective preference tests for sound fields in concert halls simulated by the aid of a computer. *J Sound Vib.* 1979;65:229–239.
11. Ando Y. Calculation of subjective preference at each seat in a concert hall. *J Acoust Soc Am.* 1983:873–887.
12. Ando Y, Okano T, Takezoe Y. The running autocorrelation function of different music signals relating to preferred temporal parameters of sound fields. *J Acoust Soc Am.* 1989;86:644–649.
13. Mouri K, Akiyama K, Ando Y. Relationship between subjective preference and the alpha-brain wave in relation to the initial time delay gap with vocal music. *J Sound Vib.* 2000;232:139–147.
14. Kato K, Ando Y. A study on the blending of vocal music with the sound field by different singing styles. *J Sound Vib.* 2002;258:463–472.
15. Kato K, Fujii K, Kawai K, Ando Y, Yano T. Blending vocal music with the sound field—the effective duration of the autocorrelation function of Western professional singing voices with different vowels and pitches. In: *Proceedings of the International Symposium on Musical Acoustics.* Nasa: ISMA; 2004.
16. Schroeder MR, Gottlob D, Siebrasse KF. Comparative study of European concert halls: correlation of subjective preference with geometric and acoustic parameters. *J Acoust Soc Am.* 1974;56:1195–1201.
17. Ando Y, Sakamoto M. Superposition of geometries of surface for desired directional reflections in a concert hall. *J Acoust Soc Am.* 1988;84:1734–1740.
18. Mosteller F. Remarks on the method of paired comparisons. III. *Psychometrika.* 1951;16:207–218.
19. Gulliksen H. A least square solution for paired comparisons with incomplete data. *Psychometrika.* 1956;21:125–134.
20. Torgerson WS. *Theory and Methods of Scaling.* New York: Wiley; 1958.
21. Ando Y. Electrical and Magnetic Responses in the Central Auditory System. In: Cariani P, ed. *Auditory and Visual Sensations.* New York: Springer-Verlag; 2009; Chapter 4.
22. Ando Y, Noson D. *Music and Concert Hall Acoustics Conference Proceedings of MCHA 1995.* London: Academic Press; 1997.
23. Suzumura Y, Sakurai M, Ando Y, et al. An evaluation of the effects of scattered reflections in a sound field. *J Sound Vib.* 2000;232:303–308.
24. Ando Y, Singh PK. A simple method of calculating individual subjective responses by paired-comparison tests 14-A. *Memoirs of Graduate School of Science and Technology, Kobe University.* 1996:57–66.

CHAPTER 9

Temporal and Spatial Features of Speech Signals

BACKGROUND

Over several decades, mainly in the context of architectural acoustics and concert hall design, we have developed a comprehensive theory of auditory signal processing that is based on two internal auditory representations, the monaural autocorrelation function (ACF) and the binaural interaural cross-correlation function (IACF).[1-4] The present work extends this auditory signal processing theory to handle speech distinctions.[5]

Features of the two correlation-based representations predict major auditory percepts associated respectively with nonspatial temporal sound qualities and nontemporal spatial attributes. Features of the monaural ACF, such as the width of the central ($W_{\phi(0)}$), zero-lag peak (listening level [LL]) ($W_{\phi(0)}$), the positions (τ_1) and relative amplitudes of positive lag peaks (ϕ_1), and the effective duration of the autocorrelation (τ_e), correspond to perceptual qualities related to loudness, timbre, pitch and pitch strength, and duration. Fig. 7.10 (upper part) schematizes the dependencies between perceptual qualities and these features and illustrates their basis in the ACF. Features of the IACF correspond to binaural perceptual attributes of binaural LL, sound direction, apparent source width (ASW), and subjective diffuseness (envelopment), as shown in Fig. 7.10 (lower part). We observed neural correlates of ACF-related percepts, which we called "temporal percepts," in electrical and magnetic neural responses over the left cerebral cortical hemisphere, whereas those of IACF-related percepts, which we called "spatial percepts," we observed over the right hemisphere.[4] Since 2004, magnetoencephalography (MEG)-evoked responses in relation to ACF factors and IACF factors have been investigated by Yoshiharu Soeta and colleagues (Chapter 5).

The correlational features of primary importance to speech recognition lie in the monaural ACF. For this reason, spatial attributes will be ignored except for in section "Effects of Spatial Factors on Speech Perception.". Although the ACF contains the same information as the power spectrum, some auditory percepts, such as the pitch of the fundamental of harmonic complex tones, and timbre, have simpler and more obvious corresponding time-lag domain autocorrelation features than their counterparts in frequency-domain power spectrum features. Early analyses of running speech, such as correlatograms produced by Bennett[6] and Biddulph,[7,8] displayed running short-term autocorrelations.

As was noted then, as with voice pitch, autocorrelation-based temporal features of speech sounds may capture aspects of the sound patterns of speech elements that underlie phonetic contrasts that are not readily apparent in traditional frequency domain representations.

Autocorrelation-like temporal representations of sound exist in early auditory stations. These population-wide, mass-statistical distributions of spike timing patterns, called population-interval distributions[9,10] or summary autocorrelations,[11] resemble half-wave rectified versions of stimulus ACFs and carry periodicity and spectrum information for periodicities up to the limit of functionally significant neural phase locking, approximately 4—5 kHz. Features of population-interval distributions can account, with precision and robustness, for many diverse aspects of pitch perception and the timbral qualities of vowels.[12]

In the monaural, ACF representation used here, pitch is represented by the factor τ_1, and pitch strength is described by $\phi_1{}^{2,13}$ in the normalized ACF (Figs. 7.5 and 7.10). Loudness has been described by two factors of the ACF: LL and effective duration, τ_e.[14-16] LL is derived from the absolute magnitude of the zero-lag peak in the unnormalized ACF. Effective duration, τ_e, is the time delay at which the ACF has declined to 10% of its zero-lag, maximal value. Effective duration quantifies persistence of periodic structure in time.

Timbre is commonly defined as consisting of sound qualities that can differ when other percepts, pitch, loudness, duration, and spatial attributes, are held constant. Timbre was considered as a multidimensional perceptual attribute; however, that timbral distinctions can be produced both by changes in the spectra of stationary sounds and by transient fluxes in amplitude, frequency, and phase, it is represented by a single factor, $W_{\phi(0)}$. In terms of phonetic contrasts, vowels are examples of the former, whereas consonants are examples of the latter. In fact, some aspects of timbre are well described by the single factor, $W_{\phi(0)}$,[17] which is a measure of the relative width of the zero-lag ACF peak that corresponds to spectral tilt.

Source Signals and Methods
Stimuli
Stimuli consisted of five isolated vowels, nine CV syllables, three greeting words, and a running speech phrase from a Japanese poem. The major Japanese syllables consist of a matrix with five vowels, /a/, /i/, /u/, /e/, and /o/, and nine consonants, k, s, t, n, h, m, y, r, and w. The stimulus set for this study consisted of the five vowels and nine consonant-vowel (CV) pairs that consisted of one of the nine consonants, followed by /a/.

Three morning greeting words were chosen that begin daily conversations in three languages. The most often used greeting words in Japanese are "Ohayou-gozaimasu," in English, these are "good morning," and in German, "Guten Morgen," which were spoken by the author.

Running speech was analyzed using a phrase selected from the Japanese poem "Kane ga narunari" composed by Meiji-era Haiku poet Shiki Masaoka. The full poem text "Kaki kueba kane ga narunari horyuji" roughly translates that, while tasting a persimmon, the poet happened to hear a reverberant sound from a bell of the ancient Buddhist temple Horyuji. To study the variability of these speech signals and the robustness of the analysis, the vowels and the poem segment were recorded in four separate sessions.

Recording methods
Vowels, syllables, and phrases were spoken by the author and were recorded by an electret-condenser microphone (SONY, ECM-MS907) situated at approximately 10 cm from the mouth. Signals were fed into a computer, 16-bit sampled at 16 kHz, and passed through an A-weighting network that approximates the effects of sound transmission through the ear.[1,2]

Autocorrelation analysis
Short-term running ACFs were then computed. The running unnormalized ACF is given by Eq. (3.5).

According to previous results in modeling the running LL, the recommended signal duration $(2T)_r$ to

be analyzed was found to be approximately given by the relation[18]

$$(2T)_r \sim 30(\tau_e)_{min} \qquad (9.1)$$

where $(\tau_e)_{min}$ is the minimum value of the effective duration of the running ACF defined previously. For CV syllables, this value was $(\tau_e)_{min} \sim 1.81$ ms (Table 9.2), and for continuous speech it was $(\tau_e)_{min} \sim 1.56$ ms. A uniform value of $(2T)_r$ was consequently selected for use with both stimulus types (i.e., just shorter than 54 ms or 49 ms), so that more detailed data could be obtained when we select $(2T)_r \sim 40$ ms.

In some cases, if there are silent intervals in running speech, such as those produced by pause or closure parts or unvoiced plosives, the normalized running ACF can produce spurious results. In these situations, adding a very low level of white noise, say—50 dB in reference to the peak sound pressure level of speech signals, can improve the robustness of the analysis of the running ACF.

Five independent temporal factors were then extracted from each 2T-duration frame of the running ACF, defined in Chapter 3 of section "Analysis of Sound Signals."

SINGLE VOWELS AND SINGLE CONSONANT-VOWEL SYLLABLES
Single Vowels
Japanese vowels /a/, /i/, /u/, /e/, and /o/ were separately pronounced by the author. The five temporal factors were computed from the running ACF as a function of time, using an integration window of 40 ms and a running frame stepsize of 20 ms. The trajectories of factor values for the vowel /a/ are shown in Fig. 9.1A—C.

In all figures in this volume, arrows signify the time at which the minimum value of the effective duration, $(\tau_e)_{min}$, of each vowel or CV syllable was observed. This time point of minimal effective duration indicates the time at which the signal is changing most rapidly.

Remarkably, the times of the first minimal effective duration $(\tau_e)_{min}$ were always observed in the initial part of all the vowels. Thus the trajectory of effective duration can indicate a specific time that separates the initial, variable-F0 vowel segment from the later, stable-F0 part. Even if a vowel is pronounced with short duration, it can nevertheless be perceived as the same vowel, suggesting that the initial segment bounded by $(\tau_e)_{min}$ contains information sufficient for recognition.

Factor τ_1 corresponds to voice pitch period, 1/F0, and thus pitch frequency is $1/\tau_1$ Hz. Voice pitch was

TABLE 9.1

Five factors extracted from the running ACF ($W_\phi(0)$, τ_1, ϕ_1, $\Delta\phi_1/\Delta t^3$, τ_e, excluding LL) for five Japanese vowels, which were extracted from the ACF of speech signals after passing through the A-weighting network in condition of the speech signal between the maximum pressure levels, $(LL)_{max}$, and $[(LL)_{max}$ -30 dB]. Signals are obtained throughout with $2T = 40$ ms and the running step of 20 ms.

Factor	(a) Raising factors at $(\tau_e)_{min}$ observed				
	/a/	/i/	/u/	/e/	/o/
$W_\phi(0)$ [ms]	0.27	0.21	0.91	0.34	0.32
τ_1^2[ms]	34.8	2.78	2.87	1.72	38.8
Pitch [Hz]	28.7	359	348	581	25.7
ϕ_1	0.21	0.23	0.23	0.32	0.38
$(\tau_e)_{min}$ [ms]	2.23	9.0	11.1	3.73	2.35

Factor	(b) Beginning vowels with almost a stationary pitch τ_1 after n-th frame				
	/a/	/i/	/u/	/e/	/o/
$W_\phi(0)$	0.38	0.89	0.45	0.37	0.52
$W_{\phi(0)min}$ [ms]	0.44	1.04	0.90	0.52	0.59
τ_1 [ms] Pitch	6.75	5.53	6.21	6.82	6.43
[Hz]	148	180	161	146	155
ϕ_1	0.71	0.84	0.73	0.64	0.82
$\Delta\phi_1/\Delta t^3$	0.48	0.80	0.82	0.42	0.70
τ_e [ms]	35.4	14.9	33.3	3.64	39.9
$(n)^2$	(3)	(6)	(3)	(3)	(2)

[1]Pitch frequency is given by $1/\tau_1$ (Hz).[5]

[2]n: Gap (n frames) between $(\tau_e)_{min}$ and beginning of vowels with almost a stationary pitch τ_1 (ms). For example, the gap n = 1 means 20 ms.

[3]An additional factor of single syllables, $\Delta\phi_1/\Delta t$ as a pitch development speed in where $\Delta t = 100$ ms (Figure 9.4)

TABLE 9.2
Five factors extracted from the running ACF ($W_{\phi(0)}$, τ_1, ϕ_1, $\Delta\phi_1/\Delta t^3$, τ_e, excluding LL) for ten single syllables V only and CV with different consonants and an identical vowel /a/

(A) CONSONANTS PARTS

Factors	Raising factors at $(\tau_e)_{min}$ observed for vowel /a/ only, and with CV different consonants									
	a	ka	sa	ta	na	ha	ma	ya	ra	wa
$W_{\phi(0)}$ [ms]	0.37	0.06	0.42	0.39	0.34	0.35	1.08	1.19	0.42	0.65
τ_1 [ms] Pitch	38.1	0.18	8.25	1.62	7.93	39.8	4.31	36.6	8.31	2.50
[Hz]	26.2	555	121	617	126	25.1	232	27.3	120	400
ϕ_1	0.38	0.20	0.85	0.32	0.65	0.59	0.67	0.57	0.73	0.39
$(\tau_e)_{min}$ [ms]	4.74	1.99	3.65	2.56	4.25	1.81	19.5	4.54	4.88	7.50

(B) VOWELS PART

Factors	Beginning of vowels with almost a stationary pitch τ_1									
	a	ka	sa	ta	na	ha	ma	ya	ra	wa
$W_{\phi(0)}$ [ms]	0.40	0.26	0.42	0.39	1.18	0.34	0.36	1.19	1.19	0.42
$W_{\phi(0)min}$ [ms]	0.44	0.39	0.42	0.43	1.19	0.39	1.21	1.35	1.35	1.56
τ_1 [ms] Pitch	8.93	8.81	8.25	7.81	7.93	8.00	7.68	8.25	8.31	7.31
[Hz]	111	113	121	128	126	125	130	121	120	136
ϕ_1	0.73	0.74	0.85	0.77	0.65	0.67	0.81	0.70	0.73	0.89
$\Delta\phi_1/\Delta t$	0.20	0.17	0.69	0.70	0.08	0.26	0.24	0.27	0.10	0.14
τ_e [ms]	45.2	5.72	3.65	80.7	4.25	631	7.0	29.5	4.88	166
(n)[1]	(12)	(2)	(1)	(3)	(0)	(1)	(6)	(2)	(0)	(5)

[1]n: Gap (n frames) between $(\tau_e)_{min}$ and beginning of vowels with a stationary pitch τ_1 (ms).

constant as a function of time after $(\tau_e)_{min}$ (i.e., during the stationary part of the vowel). In the initial, variable segment before $(\tau_e)_{min}$, τ_1 showed large and/or fluctuating values.

Factor $W_{\phi(0)}$ also showed fluctuations just after $(\tau_e)_{min}$. A larger value of $W_{\phi(0)}$ signifies relatively more energy in low-frequency components, whereas a small value signifies relatively more energy in higher frequencies. A relationship between fluctuations of $W_{\phi(0)}$ and pitch strength ϕ_1 was apparent: pitch strength lessened when the spectrum tilted towards higher frequencies.

Table 9.1A and B list the values of four factors for all of the vowels at two time points, at $(\tau_e)_{min}$ and after $(\tau_e)_{min}$, respectively. The first time point $(\tau_e)_{min}$ lies at the end of the initial variable segment, and the second time point lies in the quasistationary segment that follows $(\tau_e)_{min}$, where τ_1 values have stabilized. Maximal values of $W_{\phi(0)}$ in parts of vowels, namely $W_{\phi(0)max}$, are listed in Table 9.1B.

Single Consonant-Vowel Syllables

Similar to the earlier study, nine Japanese single CV syllables were separately pronounced by the author. Each CV syllable consisted of an initial consonant C (/k/, /s/, /t/, /n/, /h/, /m/, /y/, /r/, or /w/) coarticulated with the vowel V /a/. For each syllable, the five temporal ACF factors were extracted from the running ACF. All of the syllables had a single peak LL in the range within 30 dB of the maximum level.

The time courses of $W_{\phi(0)}$ for the nine CV syllables and the isolated vowel /a/ are shown in Fig. 9.2A–C. Magnitudes and peak durations of $W_{\phi(0)}$ differed greatly across CV syllables. Remarkably, there are obvious differences between the CV syllables as to the time courses and magnitude trajectories over which $W_{\phi(0)}$ converges to the steady-state value of approximately 0.4 ms for the vowel /a/.

Dynamic development of factor ϕ_1 (voice pitch strength) continued from the beginning of the syllable, through the offsets of consonants and the onset of

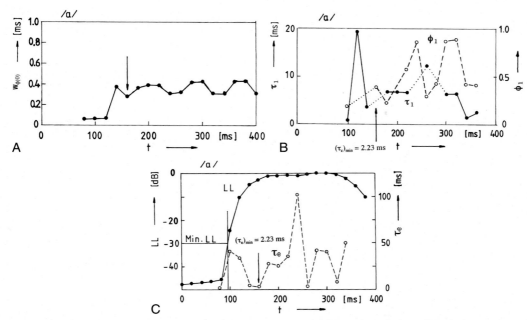

FIG. 9.1 Examples of the five temporal factors. $W_{\phi(0)}$, τ_1, ϕ_1, and τ_e, in addition to listening level (LL), as a function of time t (frame) for vowel /a/. The distance calculated due to each of the six factors is listed in Table 9.6. These have been obtained in the range LL > −30 dB in reference to the maximum level (LL = 0 dB). The sampling frequency is 16.0 kHz, 2T = 40 ms, and the running step is 20 ms, so that the running step of a single frame is 20 ms. *Arrows* indicate the time when the minimum effective duration of the autocorrelation function (ACF), $(\tau_e)_{min}$ is observed. (A) Representative of spectrum, factor $W_{\phi(0)}$. (B) Pitch-related factors, τ_1 (pitch period) and ϕ_1 (pitch strength). (C) LL and effective duration of ACF envelope (τ_e).

vowels to the end of the syllable (Fig. 9.3). Factors τ_1 (pitch period) are also plotted in the figure. There is continuity of its time course from C to V, so that four factors just after $(\tau_e)_{min}$ may provide information related to both consonant and vowel. The time of $(\tau_e)_{min}$, indicated in the plots by arrows, was always observed in initial parts of each CV, such that it can mark the boundary between consonant and vowel segments.

Table 9.2 lists five factors obtained at the time of minimal effective duration $(\tau_e)_{min}$, generally at CV boundaries (see Table 9.2A), and at times in the quasistationary vowel segment that followed (see Table 9.2B).

Results of pitch-related factors τ_1 and ϕ_1 for CV syllables (sa, ta, ha, ma, ya, ra) are gathered in Fig. 9.3A–F. Common behavior of factors τ_1 and ϕ_1 are schematically illustrated by the heavy lines in Fig. 9.4. As in the figure, the symbol n in this figure is a gap that represents the number of 20-ms frame steps between the time at $(\tau_e)_{min}$ and the time in the following vowel segment where voice pitch strength is higher and the pitch period τ_1 has stabilized. These frame numbers (n) are reported

in the bottom row of Table 9.2, showing that $(\tau_e)_{min}$ is observed before continuous pitch of the vowel.

It was found that around the time of minimal effective duration $(\tau_e)_{min}$, the speed of development of pitch strength given by $\Delta\phi_1/\Delta t$ varied according to syllables (Tables 9.1–9.3). For example, sa and ta produce greater values of $\Delta\phi_1/\Delta t$ than any other syllables.

In terms of musical timbre, $\Delta\phi_1/\Delta t$ provides a voice-pitch-based analogue of attack-decay intensity dynamics, whereas $W_{\phi(0)}$ provides an analogue of gross spectral energy distribution.[19] In the context of speech, this pitch onset dynamic might depend on the speed of pronunciation, especially at the time point of voicing onset of whole CV syllables.

As mentioned previously (see Fig. 9.4), after the n-th frame after $(\tau_e)_{min}$ and at beginnings of vowel steady-state segments, observed values for the pitch-related factor τ_1 were almost constant, signifying a strong continuous pitch that is reflected in accompanying large values for the factor $\phi_1 \sim 0.8$ (voice pitch strength).

Variations were observed for voice pitches $1/\tau_1$ of different vowels and for the same vowel /a/ following

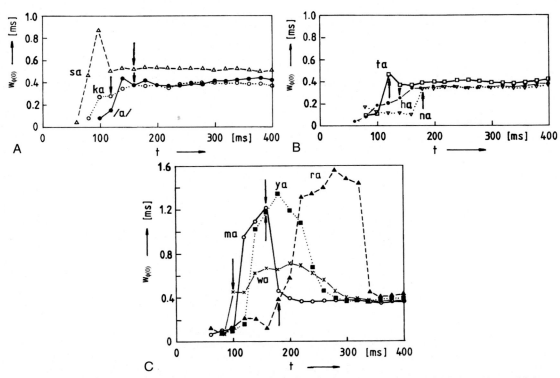

FIG. 9.2 Examples of factor $W_{\phi(0)}$ of the analyzed single consonant-vowel syllables. (A) Factors $W_{\phi(0)}$ of /a/, ka, and sa. (B) Factors $W_{\phi(0)}$ of ta, na, and ha. (C) Factors $W_{\phi(0)}$ of ma, ya, ra, and wa. Large values of $W_{\phi(0)}$ continued over the vowel parts, for example, ma, ya, ra, and wa.

different initial consonants for the same speaker. It is interesting to point out that fluctuation of pitch frequencies of single vowels was relatively widely distributed, ranging from 140 to 276 Hz, whereas those for vowel segments in the CV syllables were more narrowly distributed, ranging from 102 to 134 Hz. In addition, the factors $W_{\phi(0)}$, τ_1, and ϕ_1 similarly fluctuated more (e.g. Fig. 9.1 for /a/) for the five single isolated vowels than for vowels following consonants (CV syllables), which were, in comparison, rather stable (Figs. 9.2 and 9.3). On the contrary, pitch frequencies $(1/\tau_1)$ of single syllables observed just at time $(\tau_e)_{min}$ were widely distributed over a range of 25 to 617 Hz (Table 9.2A).

Vowel and Syllable Identification

To demonstrate potential usefulness of ACF-derived features, discrimination of isolated vowels and CV syllables using three selected ACF factors was shown. The three factors used were (1) $W_{\phi(0)max}$ (spectral tilt), (2) τ_1 (voice pitch period) obtained in the quasistationary

segment after $(\tau_e)_{min}$, and (3) $\Delta\phi_1/\Delta t$ (speed of pitch strength onset).

Because of the variability of the vowel features, four additional separate recordings of each of the five spoken vowels were made by the author. As demonstrated for vowel /a/, similar results for $\Delta\phi_1/\Delta t$ were obtained across the five recording sessions (Fig. 9.5 for a single session), and the same was true for the other vowels. The figure also includes results from the first session that were listed in Table 9.1. We can see that pitch frequencies are widely scattered for each vowel.

All factors, including $\Delta\phi_1/\Delta t$, are listed in Table 9.1B, to provide an indication of the variability of each analyzed factor. The values of factor $W_{\phi(0)max}$ listed in the tables show similarities across sessions and obvious differences across vowels. This factor therefore appears to be potentially effective for distinguishing vowels.

Vowels and CV syllables were mapped in a three-dimensional feature space (Fig. 9.6A, B) using factors $W_{\phi(0)max}$ (spectral tilt), $\Delta\phi_1/\Delta t$ (speed of pitch onset), and $1/\tau_1$ (pitch frequency). The vowels and CV syllables

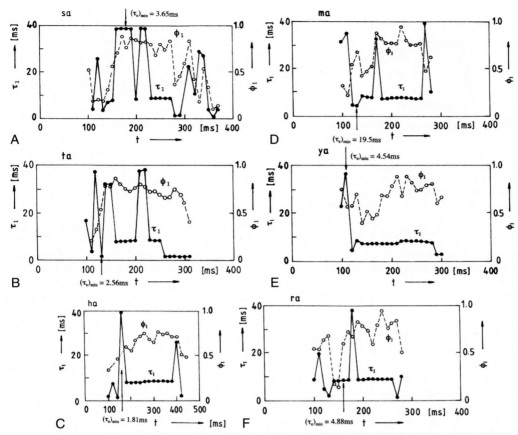

FIG. 9.3 Examples of pitch-related factors τ_1 (pitch period) and ϕ_1 (pitch strength) as a function of time (frame) of the analyzed single consonant-vowel syllables. *Arrows* indicate the time when the minimum effective duration of the autocorrelation function, $(\tau_e)_{min}$ is observed (i.e., at the beginning of syllables). (A) sa. (B) ta. (C) ha. (D) ma. (E) ya. (F) ra.

appear to be well separated on this basis, even though only three ACF factors are used. It is worth noting that the pitch frequencies of these CV syllables were similar around 200 Hz due to a combination of different syllables accompanying the vowel /a/, as shown in Fig. 9.6B.

CONTINUOUS SPEECH

Five factors are shown as a function of time, respectively for greeting words "good morning" in Japanese, English, and German, which were extracted from the running ACF, and a Japanese HAIKU phrase with and without noise conditions. Results show that segmentation of "actual syllables" has been performed by five factors extracted from the ACF, including minimum effective duration of ACF envelope, $(\tau_e)_{min}$. The

significant factors for identifying each syllable in continuous speech signals are: (1) maximum value of $W_{\phi(0)}$, (2) maximum value of sound level, LL_{max}, (3) syllable duration defined by -3 dB from LL_{max}, (4) pitch period, τ_1, and (5) the factor $\Delta\phi_1/\Delta t$.

We found that effects of added weak noise were not critical due to the fact that ACF is a correlation process. To simply demonstrate how useful the ACF factors are, three factors ($W_{\phi(0)max}$, pitch frequency, and $\Delta\phi_1/\Delta t$) of actual segmented syllables were selected and mapped in three-dimensional space.

Background

Direct evidences support the fact that the monaural ACF was found in patterns of spike timings in the auditory nerve.[9,10,20] It is remarkable that temporal primary perceptions *(pitch, timbre, loudness, duration)* have been

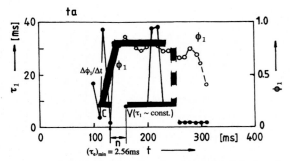

FIG. 9.4 Schematic illustration for pitch-related factors for consonant-vowel syllables. Vowel parts indicate almost constant values τ_1 (pitch period), and the factor ϕ_1 (pitch strength) approaches large values of $\phi_1 \sim 0.8$, where n is the number of the running step counted for a short gap between consonants and vowels beginning with almost constant τ_1, n = 0 to 5, as listed in Table 9.3B. $\Delta\phi_1/\Delta t$ is a slope of factor ϕ_1 obtained from consonants to vowels as a speed of pitch development. The values of $\Delta\phi_1/\Delta t$ ($\Delta t = 100$ ms) measured are indicated in Tables 9.2–9.4. A large value of $\Delta\phi_1/\Delta t$ signifies a high speed in pronouncing part C, such as /sa/ and /ta/.

directly described by the five temporal factors extracted from the short-term running ACF.[4,15] This supports previous studies by Licklider.[21,22] As discussed earlier, an initial study on identification of single syllables has been performed by use of five temporal factors extracted from the running ACF for each frame of a monaural signal.

We reconfirm the five factors that are:
1. Sound pressure level, given by factor LL;
2. Width of amplitude $\phi(\tau)$ around the origin of delay time defined at a value of 0.5, factor $W_{\phi(0)}$ *(relative weight of low- and high-frequency components)*;
3. Factor τ_1 *(period of pitch)*, the pitch frequency is simply given by $1/\tau_1$;
4. Factor ϕ_1 *(pitch strength)*, an amplitude of $\phi(\tau)$ at $\tau = \tau_1$;
5. Factor τ_e *(effective duration of the ACF: 10 percentile—amplitude delay)*, found where the minimum value $(\tau_e)_{min}$ is observed at a quick change in each syllable.

Evoked MEG evidences corresponding to temporal factors or ACF factors, LL (SPL), pitch period τ_1, pitch strength ϕ_1 and spatial factors or IACF factors of the sound field (SF), interaural cross-correlation (IACC), and τ_{IACC} were discussed in Chapter 5.

The five temporal factors analyzed for continuous speech signals and different features according to segmented syllables as a function of time (or frame period) are discussed for identification and speech recognition.

Procedure

An electret-condenser microphone (SONY, ECM-MS907) was placed at approximately 10 cm from the mouth of the speaker to receive a signal fed into a computer at a sampling frequency of 16 kHz (16 bit). The running ACF was analyzed after passing through the A-weighting network that corresponds to the ear sensitivity of listeners with normal hearing ability.

For continuous speech signals, we found here that $(\tau_e)_{min} \sim 1.56$ ms, selected 2T ~ 40 ms throughout, and the running step was 20 ms (frame time period) which was similar to the previous section with single syllables. Thus 2T was selected a little bit faster than what was calculated by Eq. (9.1), so that 2T ~ 40 ms.

Results of Five Temporal Factors
Analysis of three greetings in three languages
Common spoken greeting phrases in three languages were analyzed: "Ohayou-gozaimasu" (Japanese), "Good morning" (English), and "Guten Morgen" (German). Temporal factors were extracted from the running ACF, as listed in Table 9.4A–C. Running values for spectral tilt $W_{\phi(0)}$ and the two pitch-related factors τ_1 and ϕ_1 are plotted in Fig. 9.7A–C. As in the previous examples, segmentation of actual syllables could be easily performed by all five factors as a function of time. Arrows indicate the times of minimum effective duration $(\tau_e)_{min}$; these particularly were observed at syllable onsets and/or near sound pressure level maxima.

The three-dimensional mapping of the selected features ($W_{\phi(0)max}$, pitch frequency ($1/\tau_1$) and $\Delta\phi_1/\Delta t$) of each syllable is shown in Fig. 9.7D. Even though only a single trial for the different languages was performed, each syllable is clearly separated, and thus identification on the basis of these features appears plausible.

Japanese Haiku phrase
The phrase, "kane ga narunari," of a Meiji-era poem by Shiki Masaoka was spoken by the author twice in quiet and twice in noise. ACF-derived factors for two spoken repetitions of the phrase in quiet are listed in Table 9.5A and B. Fig. 9.8A and B show the time courses of five temporal factors, providing the observed numerical values at different time points. At the upper part of the figures, vertical lines without arrows indicate segmentation of "actual" syllables shown at the top of the plots. These lines have been determined by
1. Local dip of factor LL;
2. Each syllable should include one or two effective duration minima $(\tau_e)_{min}$, which reflect the times when sound patterns are most rapidly changing. Here vertical lines with arrows signify transition

TABLE 9.3

Five factors extracted from the running ACF ($W_{\phi(0)}$, τ_1, ϕ_1, $\Delta\phi_1/\Delta t$, τ_e, excluding LL) and the additional factor $\Delta\phi_1/\Delta t$ in beginning of vowels with a stationary pitch τ_1 after $(\tau_e)_{min}$, which are extracted from five Japanese vowels with four sessions pronounced by the author

(a) Session 1

Factor	/a/	/i/	/u/	/e/	/o/
$W_{\phi(0)}$	0.41	1.00	0.90	0.32	0.54
$W_{\phi(0)max}$ [ms]	0.41	1.09	0.90	0.47	0.56
τ_1 [ms] Pitch[1]	6.18	6.18	5.62	5.68	5.68
[Hz]	161	161	177	176	176
ϕ_1	0.65	0.62	0.87	0.67	0.79
$\Delta\phi_1/\Delta t$	0.45	0.51	1.08	0.55	0.80
τ_e [ms]	23.5	24.8	91.8	52.3	159

(b) Session 2

Factor	/a/	/i/	/u/	/e/	/o/
$W_{\phi(0)}$	0.36	0.93	0.49	0.27	0.46
$W_{\phi(0)max}$ [ms]	0.42	1.06	0.89	0.61	0.58
τ_1 [ms] Pitch	5.37	5.87	5.50	5.25	5.37
[Hz]	176	170	181	190	186
ϕ_1	0.90	0.64	0.68	0.80	0.75
$\Delta\phi_1/\Delta t$	0.53	0.45	0.83	0.46	0.78
τ_e [ms]	210	25.4	31.2	6.61	7.59

(c) Session 3

Factor	/a/	/i/	/u/	/e/	/o/
$W_{\phi(0)}$	0.37	0.98	0.88	0.46	0.48
$W_{\phi(0)max}$ [ms]	0.42	0.98	0.95	0.58	0.55
τ_1 [ms] Pitch[1]	5.62	6.06	5.81	6.12	5.56
[Hz]	177	165	172	163	179
ϕ_1	0.84	0.73	0.87	0.66	0.73
$\Delta\phi_1/\Delta t$	0.44	0.75	1.08	0.60	0.70
τ_e [ms]	86.0	33.1	50.1	49.6	15.7

(d) Session 4

Factor	/a/	/i/	/u/	/e/	/o/
$W_{\phi(0)}$	0.38	0.97	0.80	0.57	0.57
$W_{\phi(0)max}$ [ms]	0.40	1.04	0.93	0.57	0.58
τ_1 [ms] Pitch	5.93	6.00	5.37	6.00	6.50
[Hz]	168	166	186	166	153
ϕ_1	0.79	0.81	0.87	0.68	0.64
$\Delta\phi_1/\Delta t$	0.55	0.98	0.55	0.60	0.60
τ_e [ms]	49.4	47.8	17.9	42.5	2.99

[1]The pitch frequency is given by $1/\tau_1$ (Hz), where the value of τ_1 is selected at the value of ϕ_1 exceeding 0.60.

times that are observed in initial segments and/or near sound pressure level maxima;

3. Linear regression from several values of factor $\Delta\phi_1/\Delta t$, which reflects the rate of change of voice pitch strength over a 100-ms timespan.
4. Time at which sudden changes of factor τ_1 occur and;
5. Time at which sudden changes of factor $W_{\phi(0)}$ occur. A larger value of $W_{\phi(0)}$, signifies relatively greater energy in low-frequency registers, whereas a smaller value reflects relatively more energy in higher frequencies.

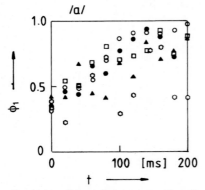

FIG. 9.5 Examples of the value $\Delta\phi_1$ as a function of time (frame) obtaining the value of $\Delta\phi_1/\Delta t$ for vowel /a/. Different symbols indicate values from five sessions.

These parameters permit a computer program to automatically segment the syllables.

To test the robustness of segmentation using ACF-based features, two additional recordings of the spoken Haiku phrase were made with noise added from an FM radio mistuned without selecting any station. The noise levels were approximately -52 dB in reference to LL_{max}. In such weak noise conditions (-51 dB), results of each factor were similar (Table 9.2A−D and Fig. 9.9A−E). However, we recommend further studies in noise conditions ($-10...-45$ dB).

The values for ACF-derived factors in noise are given in Table 9.5C and D. As with other stimuli, the values reported in the tables were obtained just after $(\tau_e)_{min}$ in the quasistationary portion of each vowel, where the pitch period (τ_1) is most stable. In addition to the factors analyzed in the previous examples, such as $(\tau_e)_{min}$ (minimum effective duration), τ_1 (pitch frequency), ϕ_1 (pitch strength), and $\Delta\phi_1/\Delta t$ (pitch strength rate of change), several other measures were also considered. The additional analyzed factors were:

1. LL relative to maximum level LL_{max},
2. Duration (n) of near-maximal LL defined by the number of 20 ms frames for which the signal level was within 3 dB of its maximum ($LL_{max} = 0$ and -3 dB),
3. Pitch frequency ($1/\tau_1$)

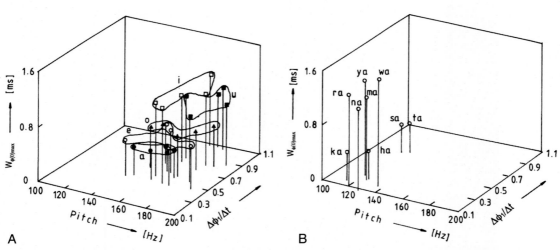

FIG. 9.6 Simple mapping in three-dimensional space. Selected factors were a representation of spectrum $W_{\phi(0)max}$, pitch frequency ($1/\tau_1$), and pitch development $\Delta\phi_1/\Delta t$, which are observed around $(\tau_e)_{min}$, as shown in Fig. 9.6 (see Tables 9.2−9.4). (A) Five vowels (five plots in total). (B) Consonant-vowel syllables.

TABLE 9.4

(a) Factors analyzed for Japanese greeting in the morning: "Ohayou gozaimasu"

Factors	Beginning of vowels with almost a stationary pitch τ_1					
	Oha	yo	go	zai	ma	su
Relative LL$_{max}$ [dB]	−29.4	−25.4	−32.3	−26.3	−32.1	−42.5
Duration for −3 dB [n][1]	7	6	2	3	3	3
$(\tau_e)_{mix}$	2.49	2.94	2.52	3.81	2.76	4.70
τ_1 [ms]	8.31	7.87	8.25	8.63	9.56	0.18
Pitch[2] [Hz]	120	127	121	115	104	5555
ϕ_1	0.63	0.72	0.73	0.65	0.63	0.62
$\Delta\phi_1/\Delta t$[3] Average: 4.4	3.20	2.85	4.15	6.20	6.10	4.00
$W_{\phi(0)max}$ [ms]	0.82	0.56	1.14	0.57	1.11	1.42

(B) FACTORS ANALYZED FOR CONTINUOUS SPEECH OF "GOOD MORNING"

Factors	Beginning of vowels with almost a stationary pitch τ_1			
	Good	mor	ni	ng
Relative LL$_{max}$ [dB]	−27.0	−22.8	−31.0	−35.7
Duration for −3 dB [n][1]	3	5	6	2
$(\tau_e)_{min}$[2]	7.63	4.10	5.43	2.56
τ_1 [ms]	5.93	7.18	7.25	6.81
Pitch[3] [Hz]	168	139	137	146
ϕ_1	0.73	0.79	0.86	0.86
$\Delta\phi_1/\Delta t$[4]	7.00	8.00	10.5	2.10
Average: 6.9 $W_{\phi(0)max}$ [ms]	1.23	1.21	1.12	1.03

(C) FACTORS ANALYZED FOR GERMAN GREETING IN THE MORNING: "GUTEN MORGEN"

Factors	Beginning of vowels with almost a stationary pitch τ_1			
	Gu	ten	Mor	gen
Relative LL$_{max}$ [dB]	−33.0	−31.8	−26.9	−29.5
Duration for −3 dB [n][1]	3	3	6	4
$(\tau_e)_{min}$[2]	7.45	4.07	4.35	6.12
τ_1 [ms]	6.93	8.75	7.50	8.18
Pitch[3] [Hz]	144	114	133	122
ϕ_1	0.80	0.62	0.68	0.72
$\Delta\phi_1/\Delta t$[4] Average: 6.08	5.80	7.70	4.05	6.80
$W_{\phi(0)max}$ [ms]	1.05	1.14	1.34	1.20

[1]Duration measured by the number of frames (n) defined by −3 dB from LL$_{max}$, unit of n is 20 ms.
[2]Pitch frequency is given by $1/\tau_1$ (Hz) measured in a condition of $\phi_1 > 0.65$.
[3]$\Delta\phi_1/\Delta t$ is a speed of pitch development measured for each syllable for
$\Delta t = 100$ ms.

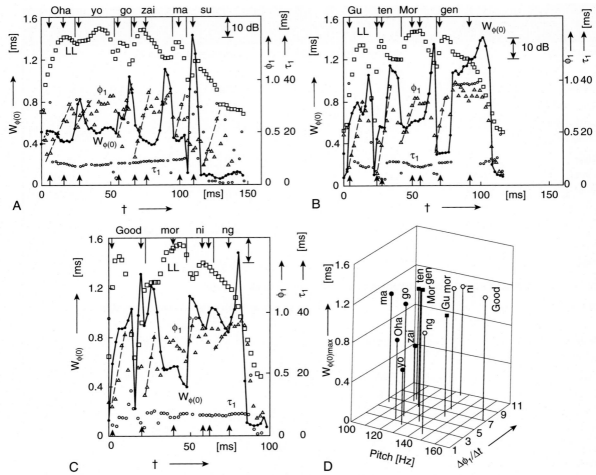

FIG. 9.7 Five temporal factors as a function of time extracted from the running autocorrelation function (ACF) of a continuous speech signal. The times of minimum value, $(\tau_\varepsilon)_{min}$, observed are indicated by arrows. : Relative sound pressure level or listening level (LL); ●: $W_{\phi(0)}$; ○: Pitch period τ_1; △: Pitch strength ϕ_1, and *dashed line*: development of the pitch strength attaining $\Delta\phi_1/\Delta t$. (A) Japanese greeting in the morning: "Ohayou gozaimasu," single trail. (B) German greeting in the morning: "Guten Morgen," single trail. (C) English greeting in the morning: "Good morning," single trail. (D) Three-dimensional mapping of each syllable in continuous speech signals of greeting in Japanese ("Ohayou gozaimasu."), English ("Good morning"), and German ("Guten Morgen"), single trail. Selected ACF factors are (1) the maximum value of $W_{\phi(0)}$, (2) pitch ($1/\tau_1$), and (3) the factor $\Delta\phi_1/\Delta t$.

4. Normalized value of $\Delta\phi_1/\Delta t$ relative to averaged values of four trials. Note that absolute values of $\Delta\phi_1/\Delta t$ greatly depend on the speed of pronunciation and/or speaking rate, such that normalizing this parameter reduces its variability.

5. Maximum values of $W_{\phi(0)}$ in parts of each syllable, namely $W_{\phi(0)max}$.

Each of these five factors is shown in Fig. 9.9A–E. No systematic differences between quiet (−61 dB S/N re: maximum signal SPL) and noise (−52 dB S/N) conditions were seen. Thus a weak noise such as Gaussian noise produced by a mistuned FM radio has little influence on ACF-derived features used for speech recognition.

To briefly examine reliability of identification, three factors in Table 9.5 were selected (i.e., [1] $W_{\phi(0)max}$, [2] pitch frequency, and [3] $\Delta\phi_1/\Delta t$ normalized by averaged values over four trials. As in previous

TABLE 9.5
Five temporal factors and associated variables for continuous speech signals of HAIKU by Shiki Masaoka: "Kanega narunari"

(A) TRIAL 1 (WITHOUT NOISE; S/N ~ 61 MEASURED AT THE MAXIMUM SPL OF "GA")

Factors	Beginning of vowels with almost a stationary pitch τ_1						
	ka	ne	ga	na	ru	na	ri
Maximum SPL LL_{max} [dB]	−31.7	−32.4	−22.6	−30.0	−39.3	−29.1	−43.5
Duration for −3 dB [n][1]	3	6	5	4	5	3	2
Value $(\tau_e)_{min}$	2.94	31.8	5.63	2.72	4.68	4.40	4.04
τ_1 [ms]	6.62	7.12	6.68	7.56	7.68	7.93	11.0
Pitch[3] [Hz]	151	140	149	132	130	126	90
ϕ_1	0.78	0.74	0.86	0.78	0.77	0.73	0.61
$\Delta\phi_1/\Delta t$[3] Average: 3.70	4.25	2.85	2.85	2.30	3.80	4.85	5.00
Normalized[4] 3.95	4.54	3.04	3.04	2.46	4.06	5.18	5.35
$W_{\phi(0)max}$ [ms]	0.40	1.02	0.40	1.35	1.25	1.45	1.15

(B) TRIAL 2 (WITHOUT NOISE; S/N ~ 61 MEASURED AT THE MAXIMUM SPL OF "GA")

Factors	Beginning of vowels with almost a stationary pitch τ_1						
	ka	ne	ga	na	ru	na	ri
Relative LL_{max} [dB]	−36.5	−32.2	−21.6	−30.3	−37.6	−28.2	−42.4
Duration for −3 dB [n]	3	7	4	4	6	4	4
$(\tau_e)_{min}$	4.16	3.69	15.1	3.72	3.13	5.49	5.23
τ_1 [ms]	7.25	7.06	6.25	7.81	7.37	7.93	9.0
Pitch [Hz]	137	141	160	128	135	126	111
ϕ_1	0.86	0.76	0.62	0.60	0.78	0.78	0.70
$\Delta\phi_1/\Delta t$ Average: 4.13	6.50	5.10	4.50	2.20	4.25	3.05	3.35
Normalized 4.00	6.30	4.95	4.36	2.13	4.10	2.96	3.25
$W_{\phi(0)max}$ [ms]	0.43	0.99	0.40	1.25	1.18	1.12	1.20

(C) TRIAL 3 (WITH NOISE; S/N ~ 52 DB MEASURED AT THE MAXIMUM SPL OF "GA")

Factors	Beginning of vowels with almost a stationary pitch τ_1						
	ka	ne	ga	na	ru	na	ri
Relative LL_{max} [dB]	−33.6	−27.4	−19.3	−29.4	−30.4	−26.6	−41.0
Duration for −3 dB [n]	3	6	4	3	5	3	3

Continued

TABLE 9.5
Five temporal factors and associated variables for continuous speech signals of HAIKU by Shiki Masaoka: "Kanega narunari"—cont'd

(C) TRIAL 3 (WITH NOISE; S/N ∼ −52 DB MEASURED AT THE MAXIMUM SPL OF "GA")

Factors	Beginning of vowels with almost a stationary pitch τ_1						
$(\tau_e)_{min}$	6.54	4.12	4.13	3.98	3.34	13.0	4.30
τ_1 [ms]	7.43	6.75	6.50	7.93	7.68	8.12	9.43
Pitch [Hz]	134	148	153	126	130	123	106
ϕ_1	0.74	0.70	0.63	0.72	0.70	0.74	0.62
$\Delta\phi_1/\Delta t$ Average: 3.54	6.50	2.90	3.25	2.15	4.25	2.80	2.95
Normalized 3.96	7.28	3.33	3.64	2.47	4.76	3.13	3.30
$W_{\phi(0)max}$ [ms]	1.12	0.43	0.43	1.15	1.13	1.36	1.27

(D) TRIAL 4 (WITH NOISE; S/N ∼ −52 DB MEASURED AT THE MAXIMUM SPL OF "GA")

Factors	Beginning of vowels with almost a stationary pitch τ_1						
	ka	ne	ga	na	ru	na	ri
Relative LL_{max} [dB]	−29.7	−28.9	−21.1	−31.2	−32.9	−27.1	−39.1
Duration for −3 dB [n]	3	5	4	3	4	3	3
$(\tau_e)_{min}$	5.34	2.51	6.04	3.95	3.46	14.0	2.99
τ_1 [ms]	7.00	6.68	6.75	7.50	7.56	7.87	9.25
Pitch [Hz]	142	149	148	133	132	127	108
ϕ_1	0.66	0.73	0.72	0.67	0.77	0.63	0.67
$\Delta\phi_1/\Delta t$ Average: 4.60	11.6	5.50	4.35	2.75	1.90	3.15	3.00
Normalized 4.00	10.0	4.78	3.78	2.39	1.65	2.73	2.60
$W_{\phi(0)max}$ [ms]	0.55	1.06	0.47	1.29	1.23	1.34	1.25

[2] Pitch frequency is given by $1/\tau_1$ (Hz) measured in a condition of $\phi_1 > 0.60$.
[1] Duration measured by the number of frames (n) defined by −3 dB from LL_{max}, unit of n is 20 ms.
[3] $\Delta\phi_1/\Delta t$ is a speed of pitch development measured for each syllable for $\Delta t = 100$ ms.
[4] Normalized value is obtained by averaged $\Delta\phi_1/\Delta t$ of 4 trials, because we considered this value fluctuates according to speed of pronunciation.

TABLE 9.6
The distance calculated due to each of the six factors

Horizontal angle of noise disturbance	Temporal	factors		Spatial	factors	
	$D\tau_e$	$D\tau_1$	$D\phi_1$	D_{IACC}	Dw_{IACC}	$D\Phi(0)$
30°	0.420	0.164	0.442	0.248	0.052	0.064
60°	0.351	0.247	0.355	0.266	0.049	0.056
90°	0.348	0.162	0.401	0.292	0.049	0.063
120°	0.279	0.157	0.376	0.270	0.043	0.058
180°	0.383	0.171	0.494	0.247	0.071	0.074

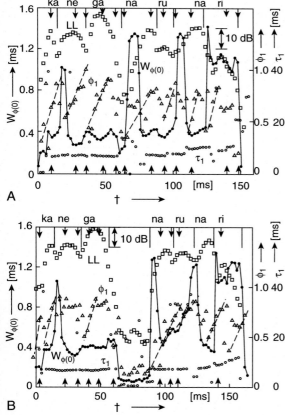

FIG. 9.8 Five temporal factors as a function of time (frame) extracted from the running autocorrelation function of continuous speech signal in Japanese, reading a part of HAIKU by Shiki Masaoka: "Kanega narunari." The times of observed minimum value, $(\tau_e)_{min}$, are indicated by arrows. (A) Trial 1. (B) Trial 2. □, Relative sound pressure level listening level; ●, $W_{\phi(0)}$; ○, pitch period τ_1; △, pitch strength ϕ_1, and dashed linear line: development of pitch strength obtained by linear regression, factor $\Delta\phi_1/\Delta t$.

examples, three-dimensional feature maps of each syllable using these selected factors are plotted in Fig. 9.10, different symbols represent values obtained in the quiet.

Conclusions

A set of independent factors were extracted from the running short-time ACFs of vowels, CV syllables, and short phrases. We found that[5]:

1. Many phonetic and syllabic distinctions are easily made with this ACF-derived feature set, which differs considerably from frequency-domain phonetic features traditionally used for speech recognition. Neural correlates for these features plausibly exist in interspike interval distributions in early auditory stations (Chapter 2).

2. Points of maximal signal change, as reflected in minimum values of the effective duration of the running ACF, $(\tau_e)_{min}$, were always observed in initial segments of single vowels and CV syllables (arrows in Figs. 9.1–9.4).

3. The factor $W_{\phi(0)}$, which reflects spectral tilt, the relative distribution of low-frequency versus high-frequency energy, is a useful measure of timbral brightness (see Figs. 9.1–9.3). Values of $W_{\phi(0)}$ obtained just after times of rapid signal change, as indicated by minimal effective duration $(\tau_e)_{min}$, may include information effective for syllable identification.

4. For the quasistationary portion of vowels, factor τ_1 was almost at constant value, signifying a strong pitch due to large values of the factor $\phi_1 \sim 0.80$ (pitch strength).

5. Dynamic behavior of both factors $W_{\phi(0)}$ and ϕ_1 marks offsets of consonants and onsets of vowels. Thus four ACF factors obtained at CV boundaries, just after $(\tau_e)_{min}$, may include information on both consonant and vowel (see Figs. 9.1–9.4).

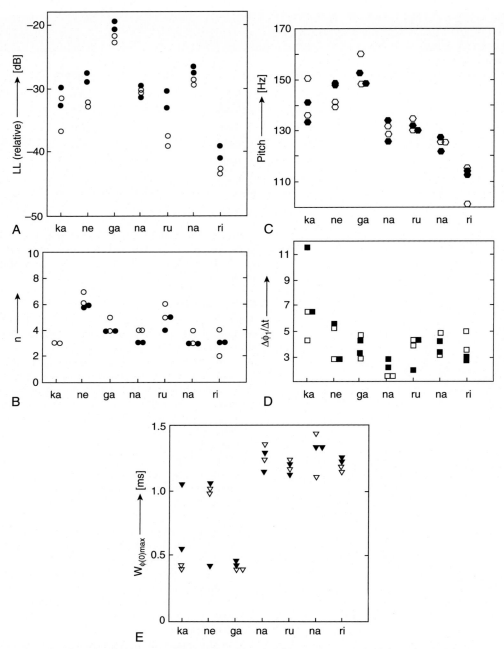

FIG. 9.9 Five factors and duration n for a continuous speech signal in Japanese, reading HAIKU: "Kanega narunari," for four trials, empty symbols without any noise, and full symbols with the weak noise.
(A) Relative listening level (LL). (B) Duration n (defined by −3 dB of LL_{max}, unit of n is 20 ms). (C) Pitch frequency $(1/\tau_1)$. (D) $\Delta\phi_1/\Delta t$. (E) $W_{\phi(0)max}$.

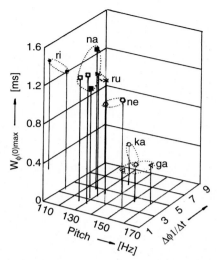

FIG. 9.10 Three-dimensional mapping of each syllable in a continuous speech signal in Japanese, reading HAIKU: "Kanega narunari," for two trials without any noise. Three selected factors are (1) maximum value of $W_{\phi(0)}$, (2) pitch (1/τ_1), and (3) factor $\Delta\phi_1/\Delta t$. Different symbols indicate results for each syllable.

6. Speed of pitch strength increase for vowels, $\Delta\phi_1/\Delta t$, may act as an additional timbral factor (see Fig. 9.4) that is analogous to attack/decay dynamics in musical contexts.
7. Satisfactory separation of CV syllables in phrases can be achieved using small numbers of ACF-derived factors, such as $W_{\phi(0)max}$, pitch frequency, and $\Delta\phi_1/\Delta t$, computed at appropriate times indicated by effective duration markers (see Fig. 9.7d).
8. The ACF-derived factors studied here appear to be robust with respect to added noise (see Fig. 9.9; Table 9.5) and bear further investigation for practical use in future automatic speech recognition (ASR).

EFFECTS OF SPATIAL FACTORS ON SPEECH PERCEPTION

We are interested in the effect of SFs on the interaction of sounds and in particular how reflections affect the quality of speech sounds. We have discussed subjective preference in terms of delay time and IACC in Chapter 8. In this section, we will discuss degradation effects of the reflections in terms of both temporal and spatial factors.

In our experiment, the loudspeaker located in front of the listener presented single syllables, and the continuous white noise as a disturbance was produced from another loudspeaker located at different horizontal angles. The three temporal factors and the sound energy were extracted from the ACF of the speech signal, and three spatial factors were extracted from the IACF for sound signals arriving at the two ear entrances. Results show that two factors, the effective duration, $(\tau_e)_{min}$, in the temporal factors extracted from the running ACF, and the W_{IACC} in the spatial factors extracted from the IACF, had significant effects on nonidentification (NI). We will attempt to calculate the identification of single syllables in the condition of noise disturbance from different directions based on the human auditory-brain model (Ando and Yamasaki, unpublished). We assume that the specialization of the human cerebral hemisphere may relate to the highly independent contribution between the spatial and temporal factors to speech identification. It is remarkable that, for example, "cocktail party effects" might well be explained by such specialization of the human brain because speech is mainly processed in the left hemisphere and independently spatial information is simultaneously processed in the right hemisphere. Based on such a model, we have described temporal and spatial percepts in Chapter 7. According to the model shown in Fig. 2.19, temporal factors associated with the left hemisphere together with the sound energy were extracted from the ACF of the sound signal arriving at one of the ear entrances. In addition, spatial factors associated with the right hemisphere were extracted from the IACF of sound signals arriving at the two ear entrances. The running ACF and the running IACF with an integration interval $2T = 30$ ms with a running step of 10 ms were analyzed.

For the identification of speech signals, the psychological distance between characteristics of single syllables due to each of the four factors was extracted from the ACF.

Here we introduce a distance between the template source signal and the SF signal. Let S_K^T be the characteristics of an isolated template-syllable K, and let S_X^{SF} be the characteristics of another syllable X in the SF; symbol T refers to the template, and SF is the SF. Let C_K^T and C_X^T be characteristics of the isolated template S_K^T of syllable K, and another syllable X in an SF S_X^{SF} be processed in the auditory-brain system. Then, the distance between S_K^T and S_X^{SF} is given by

$$d_k = D\left(S_X^{SF}, S_K^T\right) = \left|C_X^{SF}, C_K^T\right| \qquad (9.2)$$

where

$$D_{\tau e}(X, K) = \left[\sum_{i=1}^{I} \left| \log \left(\tau_e^i \left(\frac{SF}{X} \right) - \log \left(\tau_e^i \left(\frac{T}{K} \right) \right) \right) \right| \right] \Big/ I$$

$$D_{\varphi 1}(X, K) = \left[\sum_{i=1}^{I} \log \left(\varphi_1^i \left(\frac{SF}{X} \right) \right) - \log \left(\varphi_1^i \left(\frac{T}{K} \right) \right) \right] \Big/ I$$

$$D_{\tau 1}(X, K) = \left[\sum_{i=1}^{I} \log \left(\tau_1^i \left(\frac{SF}{X} \right) - \log \left(\tau_1^i \left(\frac{T}{K} \right) \right) \right) \right] \Big/ I$$

$$D_{\Phi(0)}(X, K) = \left[\sum_{i=1}^{I} \log \left(\Phi(0)^i \left(\frac{SF}{X} \right) \Big/ \Phi(0)^{max} \left(\frac{SF}{X} \right) \right) \right.$$
$$\left. - \log \left(\Phi(0)^i \left(\frac{T}{K} \right) \Big/ \Phi(0)^{max} \left(\frac{T}{K} \right) \right) \right] \Big/ I$$

$$(9.3)$$

and K and X represent the syllable number of template (T) and the syllable in an SF, respectively. In addition, i is the frame number of the running ACF and I is the total frame number.

In addition, to find effects of the different directional noises, three spatial factors are extracted from the IACF, which are associated with the right cerebral hemisphere. The distances due to the spatial factors, D_{IACC}, $D\tau_{IACC}$, and D_{WIACC}, respectively, are given by

$$D_{IACC}(X, K) = \left[\sum_{i=1}^{I} IACC^i \left(\frac{SF}{X} \right) - IACC^i \left(\frac{T}{K} \right) \right] \Big/ I$$

$$D_{\tau IACC}(X, K) = \left[\sum_{i=1}^{I} \tau_{IACC}^i \left(\frac{SF}{X} \right) - \tau_{IACC}^i \left(\frac{T}{K} \right) \right] \Big/ I$$

$$D_{W_{IACC}}(X, K) = \left[\sum_{i=1}^{I} W_{IACC}^i \left(\frac{SF}{X} \right) - W_{IACC}^i \left(\frac{T}{K} \right) \right] \Big/ I$$

$$(9.4)$$

It is worth noting that one of the temporal factors $W_{\phi(0)}$ is deeply related to one of the spatial factors, W_{IACC}.

In general, a shorter distance between the template syllable and the syllables with noise disturbance signifies higher intelligibility. According to multiple regression analysis, the NI rate of the syllable, which was not matched with the template, has been directly calculated, so that

$$NI(S_0, S_X) = S_L + S_R$$
$$= [aD\tau_e + bD\tau_1 + cD\phi_1]_L$$
$$+ [dD\Phi(0) + eDIACC + fD\tau_{IACC} + gDW_{IACC}]_R$$

$$(9.5)$$

where

$$S_L = [aD\tau_e + bD\tau_1 + cD\phi_1]_L$$

and

$$S_R = [dD\Phi(0) + eDIACC + fD\tau_{IACC} + gDW_{IACC}]_R,$$

In addition, $\Phi_P(0)$ is measured in [dBA]. Note that the LL or $\Phi(0)$ is associated with the right hemisphere (Table 6.2). Weighting coefficients a through g in Eq. (9.5) were determined by maximizing NI with experimental data.

Fourteen single syllables, /pa/, /pu/, /te/, /zo/, /bo/, /yo/, /mi/, /ne/, /kya/, /kyo/, /pya/, /gya/, /nya/, and /zya/, with 4-s intervals between syllables were presented to each subject from the frontal loudspeaker ($\xi = 0$ degree, the distance to the center of the subject's head, d = 70 cm $\pm$ 1 cm) in an anechoic chamber. The white noise applied as a disturbance was continuously produced by one of the loudspeakers located at different horizontal angles: $\xi = 30, 60, 90, 120,$ or 180 degrees (d = 70 cm). The sound pressure level measured in terms of $\Phi_p(0)$ of both speech signals and the continuous white noise was fixed 65.0 dBA at the peak level. Ten subjects participated in the experiment, each of whom was asked what syllable they heard.

For example, values of τ_e extracted from the running ACF for the signal /mi/ with and without noise ($\xi = 90$ degrees) as a function of time are shown in Fig. 9.11. The important initial half parts of the speech signal indicating $\Phi(0) < 0.5$, as shown in Fig. 9.12, of both template and test syllables with added noise were applied in computation by Eq. (9.5).

Results of the NI rate for some single syllables as a function of the horizontal angle ξ of the noise disturbance are shown in Fig. 9.13. Almost similar tendencies in the NI of these syllables were found. When the noise arrived from 30 degrees, the NI indicated maxima in the horizontal angle range tested, and when the noise was

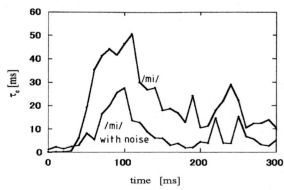

FIG. 9.11 Values of τ_e extracted from the running autocorrelation function for frontal signal /mi/ only, and /mi/ with white noise from $\xi = 90$ degrees.

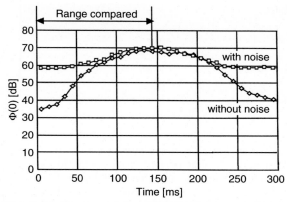

FIG. 9.12 For comparison, analyzed initial parts of a frontal single syllable with and without white noise from ξ = 90 degrees.

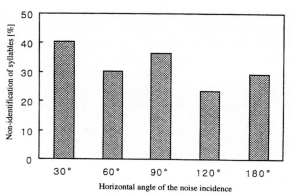

FIG. 9.14 Averaged percentile of all tested nonidentified single syllables, obtained by a listening test for different angles ξ of the white noise incidence as disturbance.

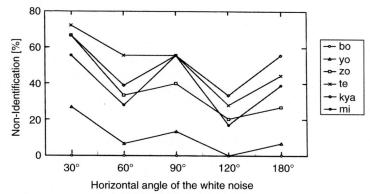

FIG. 9.13 Percentile examples of nonidentification for single syllables as a function of horizontal angle of the white noise from different horizontal angles ξ. At the horizontal angle ξ = 120 degrees, the percentage of nonidentification was minimum for single syllables.

presented from 120 degrees, NI indicated minima. The same was true for the averaged NI rate, as shown in Fig. 9.14.

Because the direct speech sound arrived at the listener from the frontal direction, the value of τ_{IACC}, being always close to zero, was an invariable. Thus this factor was eliminated from the analysis by Eq. (9.5), as listed in Table 9.7. The minima of psychological distance were always found for noise disturbance arriving from 120 degrees, so that the NIs were minimal. On the other hand, when noise disturbance arrived from 30 degrees, the distance commonly indicated maxima in six factors for all of the syllables due to τ_e.

Resulting weighting coefficients in Eq. (9.5) for the six factors are listed in Table 9.7. According to the weighting coefficients obtained here, the factors τ_e and

$W_{IACC} \sim W_{\phi(0)}$ contributed significantly to NI. For each single syllable, the relationship between the values was calculated by Eq. (9.5), and the measured values are shown in Fig. 9.15. Obviously, a linear relationship was achieved (r = 0.86, P < 0.01).

The most effective and significant temporal factor was the τ_e-value. To obtain the effects of different directions of noise disturbance, spatial factors may be taken into consideration. Conclusions are as follows:

1. Syllable NI may be calculated by both temporal factors extracted from the ACF and spatial factors extracted from the IACF.
2. Particularly in the condition of this experiment, the value of τ_e as the temporal factor is the most significant, which is similar to previous results.[14] In addition, W_{IACC}, which is deeply related to the

TABLE 9.7
Weighting coefficients in Equation (9.5) determined by the experiment

Temporal	factors		Spatial	factors	
τ_e	τ_1	ϕ_1	IACC	W_{IACC}	$\Phi(0)$
Coefficient 0.335	0.028	0.136	0.086	0.384	0.053

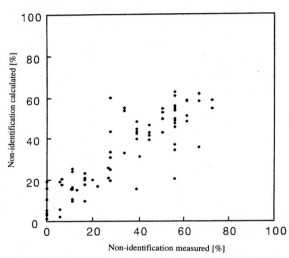

FIG. 9.15 Relationship between the calculated percentile of nonidentified syllables and those obtained by listening tests ($r = 0.86$, $P < 0.01$).

timbre factor $W_{\phi(0)}$ among the spatial factors, contributes significantly to speech identification.

FUTURE DIRECTIONS

This volume examines a neurally based theory for usefulness of autocorrelation-derived features in speech representation and recognition. As a future goal, more realistic characterization and identification of larger corpora of actual spoken syllables in terms of the selected significant ACF-derived features (e.g., $W_{\phi(0)max}$ or spectral tilt, pitch frequency, $1/\tau_1$, and $\Delta\phi_1/\Delta t$ or speed of pronunciation) including averaged or mean value of τ_e, as discussed in section "Effects of Spatial Factors on Speech Perception," could be undertaken. ASR systems based on these ACF feature primitives could also be envisioned. Given an adequate set of feature primitives, it is possible to locate phonetic classes and syllable patterns in a multidimensional feature space and to classify unknown incoming sound patterns on the basis of their distance in that percept space to the locations of known exemplars. Whether ACF-based features can better capture auditory representational invariances that underlie phonetic distinctions, such that they could be exploited by ASR systems, remains to be seen.

Automatic Speech Recognition

We propose auditory processing model–based ASR for recording of meetings and conferences and for instantaneous interpretation of different languages. As a practical example, we propose investigating possibilities for applying the auditory processing theory to cross-language speech communication and development of an even more powerful system with learning capacity and accumulation of data.

Individual Preference of Significant Temporal Factors for Hearing Aid

As discussed in Chapter 8 of section "Seat Selection Enhancing Individual Preference," we propose procedures for designing individual hearing aid systems with preferred SFs, which can be realized by adding reflections after testing significant factors in identification of syllables. Design of individually preferred monaural and binaural hearing aid systems incorporating four orthogonal factors (see Chapter 8) is well within reach. For example, adding a shorter delay time of reflection, Δt_1, than 5 ms after the direct sound, depending on individual subjective preference, results in higher perception of articulation and clarity.

Shimokura et al.[23] have obtained remarkable results from testing individuals with sensorineural hearing loss (Fig. 9.16) that indicate percent articulation increases with a longer median value of effective duration, $(\tau_e)_{med}$, of running ACF. Slowly pronounced speech without any vibrato[24] resulting in long τ_e values may increase articulation. It is noteworthy that a preferred SF results in high intelligibility, allowing adults and children alike to prefer individually designed hearing aids that support psychophysiological well-being.

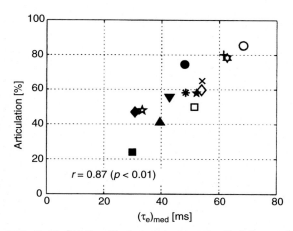

FIG. 9.16 Relationship between percent articulation and the median effective duration of running ACF, $(\tau_e)_{med}$. (From Shimokura R, Akasaka S, Nishimura T, Hosoi H. Autocorrelation factors and intelligibility of Japanese monosyllables in individuals with sensorineural hearing loss. *J Acoust Soc Am.* 2017;141:1065—1073.)

REFERENCES

1. Ando Y. *Concert Hall Acoustics.* Heidelberg: Springer-Verlag; 1985.
2. Ando Y. *Architectural Acoustics, Blending Sound Sources, Sound Fields, and Listeners.* New York: AIP Press/Springer-Verlag; 1998 [Chapters 1 through 6].
3. Ando Y. Theory of auditory temporal and spatial primary sensations. *J Temporal Des Archit Environ.* 2008;8:8—26. http://www.jtdweb.org/journal/.
4. Ando Y. *Auditory and Visual Sensations.* New York: Springer-Verlag; 2009.
5. Ando Y. Autocorrelation-based features for speech representation. *Acta Acust United Acust.* 2015;101:145—154.
6. Bennett WR. A machine for continuous display of short-term correlation—"the correlatograph". *Bell Syst Tech J.* 1953;32:1173—1185.
7. Biddulph R. Short-term autocorrelation analysis and correlatograms of spoken digits. *J Acoust Soc Am.* 1954;26:539—541.
8. Lange FH. *Correlation Techniques.* Princeton: Van Nostrand; 1967.
9. Cariani PA, Delgutte B. Neural correlates of the pitch of complex tones. I. Pitch and pitch salience. *J Neurophysiol.* 1996;76:1698—1716.
10. Cariani P. Temporal coding of periodicity pitch in the auditory system: an overview. *Neural Plast.* 1999;6:147—172.
11. Meddis R, O'Mard L. A unitary model of pitch perception. *J Acoust Soc Am.* 1997;102:1811—1820.
12. Hirahara T, Cariani P, Delgutte B. Representation of low-frequency vowel formants in the auditory nerve. In: Ainsworth WA, Greenberg S, eds. *Auditory Basis of Speech Perception.* UK: Keele; 1996:15—19, 1996:83—86. ISCA Archive http://www.isca-speech.org/archive/absp_96.
13. Inoue M, Ando Y, Taguti T. The frequency range applicable to pitch identification based upon the auto-correlation function model. *J Sound Vib.* 2001;241:105—116.
14. Ando Y, Sato S, Sakai H. Fundamental subjective attributes of sound fields based on the model of auditory-brain system. In: Sendre JJ, ed. *Computational Acoustics in Architecture.* Boston: WIT Press; 1999 [Chapter 4].
15. Ando Y, Sakai H, Sato S. Formulae describing subjective attributes for sound fields based on a model of the auditory-brain system. *J Sound Vib.* 2000;232:101—127.
16. Sato S, Kitamura T, Ando Y. Loudness of sharply (2068 dB/Octave) filtered noises in relation to the factors extracted from the autocorrelation function. *J Sound Vib.* 2002;250:47—52.
17. Hanada K, Kawai K, Ando Y. A study of the timbre of an electric guitar sound with distortion. *J S China Univ Technol (Nat Sci Ed),* In: *Proceedings of the 3rd International Symposium on Temporal Design,* Guangzhou; 2007;96—99.
18. Mouri K, Akiyama K, Ando Y. Preliminary study on recommended time duration of source signals to be analyzed, in relation to its effective duration of autocorrelation function. *J Sound Vib.* 2001;241:87—95.
19. Grey JM. Scaling of musical timbre. *J Acoust Soc Am.* 1977;61:1270—1277.
20. Secker-Walker HE, Searle CL. Time domain analysis of auditory-nerve- fiber firing rates. *J Acoust Soc Am.* 1990;88:1427—1436.
21. Licklider JCR. A duplex theory of pitch perception. *Ecxperientia.* 1951;VII:128—134.
22. Licklider JCR. Three auditory theories. In: Koch S, ed. *Psychology: A Study of a Science. Study I. Conceptual and Systematic.* New York: McGraw-Hill; 1959:41—144.
23. Shimokura R, Akasaka S, Nishimura T, Hosoi H. Autocorrelation factors and intelligibility of Japanese monosyllables in individuals with sensorineural hearing loss. *J Acoust Soc Am.* 2017;141:1065—1073.
24. Kato K, Fujii K, Kawai K, Ando Y, Yano T. Blending vocal music with the sound field—the effective duration of the autocorrelation function of Western professional singing voices with different vowels and pitches. In: *Proceedings of the International Symposium on Musical Acoustics.* Nara: ISMA 2004; 2004.

Index

Note: Page numbers followed by "f" indicate figures.